Edible Wild Plants

Effective Tips and Tricks to Procuring Nutritious and Delicious Wild Plants

(The Ultimate Guide to Foraging, Identifying, Harvesting and Cooking Essential Wild Food to Enjoy Edible Wild Plants Safely)

Chad Wells

Published By **Ryan Princeton**

Chad Wells

All Rights Reserved

Edible Wild Plants: Effective Tips and Tricks to Procuring Nutritious and Delicious Wild Plants (The Ultimate Guide to Foraging, Identifying, Harvesting and Cooking Essential Wild Food to Enjoy Edible Wild Plants Safely)

ISBN 978-1-7752672-8-7

No part of this guidebook shall be reproduced in any form without permission in writing from the publisher except in the case of brief quotations embodied in critical articles or reviews.

Legal & Disclaimer

The information contained in this ebook is not designed to replace or take the place of any form of medicine or professional medical advice. The information in this ebook has been provided for educational & entertainment purposes only.

The information contained in this book has been compiled from sources deemed reliable, and it is accurate to the best of the Author's knowledge; however, the Author cannot guarantee its accuracy and validity and cannot be held liable for any errors or omissions. Changes are periodically made to this book. You must consult your doctor or get professional medical advice before using any of the suggested remedies, techniques, or information in this book.

Upon using the information contained in this book, you agree to hold harmless the Author from and against any damages, costs, and expenses, including any legal fees potentially resulting from the application of any of the information provided by this guide. This disclaimer applies to any damages or injury caused by the use and application, whether directly or indirectly, of any advice or information presented, whether for breach of contract, tort, negligence, personal injury, criminal intent, or under any other cause of action.

You agree to accept all risks of using the information presented inside this book. You need to consult a professional medical practitioner in order to ensure you are both able and healthy enough to participate in this program.

Table Of Contents

Chapter 1: An Introduction To Edible Wild Plants 1

Chapter 2: Are Wild Edible Plants Safe To Eat? 26

Chapter 3: Harvesting Edible Wild Plants 44

Chapter 4: Being Careful When You Forage 63

Chapter 5: How To Grow Your Own Edible Wild Plants 83

Chapter 6: Common Edible Wild Plants And Their Profiles 98

Chapter 7: Common Edible Wild Plants And Their Profiles (Contd.) 137

Chapter 8: Common Poisonous Plants And Their Profiles 165

Chapter 9: Edible Wild Plants Recipes .. 174

Chapter 1: An Introduction To Edible Wild Plants

Edible wild plants, referred to as foraged food and wild substances, are the loose and ready-to-consume give up end result and greens within the wild. These vegetation increase in areas wherein humans have no longer deliberately planted and cultivated, like within the woods or the wild. Hunters and fishermen use the ones vegetation to add taste to their dishes on best activities, while exquisite people rely upon them for food.

Some wild flowers are useful in critical fitness situations, which include liver damage and most cancers. These flowers flavor suitable and may function medication and remedy critical health situations. In addition, they are rich in fibers and minerals at the side of iron, calcium, antioxidants, and zinc. This chapter data vital records approximately wild vegetation, suitable for eating flowers, identifying them, and the versions among flora, flowers, and weeds.

What Are Edible Wild Plants?

Edible flowers have a exquisite taste and function sparkling or cooked vegetables. Some are used for medicinal capabilities, which include treating liver, digestive, and respiration issues. In addition, a few flora are accurate for eyesight. They are superb stimulants and particular forms of beneficial beneficial useful resource in detoxing. People global use those vegetation for specific functions, collectively with decorations and perfumes or candles. Typically, the ones flora increase in uncultivated fields and alongside roadsides or in the wild and normally do no longer require direct planting with the aid of people. However, they're moreover grown in greenhouses and yards.

The time period "wild meals" refers to vegetation that broaden with out cultivation. Among these are match for human consumption plants with wholesome vitamins which can be utilized in vicinity of cultivated vegetation. Most wild vegetation were added from unique components of the region and expand in uncultivated areas. In many cases, wild food is better to your health than cultivated food due to the fact they lack risky chemical substances typically observed in cultivated plant life. Some wild meals plants are critical spices and veggies in cuisines. On the other hand, people have changed cultivated flowers for the vitamins due to name for. However, they lose a number of the nutrients in the technique.

Types of Edible Plants

Wild edibles are decided in maximum elements of the world. In some instances, the ones plant life also are to be had as cultivated flora in grocery stores. In maximum instances, wild appropriate for ingesting vegetation are much like the ones cultivated. They are modified from their actual form to fulfill great necessities. The variations are large. For example, cultivated pink cabbage has an first-rate flavor and coloration, but it does now not have as many nutrients as wild cabbage.

Edibles are flowers that can be eaten or used for cooking, baking, and different culinary capabilities. For example, salads normally include give up quit result, nuts, leaves, and stems of numerous plant life. Some suit to be eaten flora are proper as fundamental dishes, and others are used as candies. Plants have a combination of various parts used for cooking purposes. In addition, maximum herbs are in shape for human consumption plant life, and certain plant life are common wild edibles.

1. Fruits

Fruits are the in form for human consumption part of flowering plants. They include seeds generally used for replica. Fruits incorporate a enough range of nutrients and minerals crucial for the human body, which include calcium, vitamins A and C, potassium, phosphorus, magnesium, and so forth. Depending on their length and taste, fruits may be used as desserts or preserved to make fruit jams and pies. Their juice is used to make alcoholic drinks and jelly.

2. Nuts

Nuts are commonly enclosed in a hard and woody shell and include a unmarried seed. Most nuts may be eaten uncooked or roasted and characteristic immoderate quantities of protein and fat, which includes almonds, cashews, pecans, and pine cuts. Nuts are an extremely good supply of power and healthy vitamins and comprise biotin, fiber, magnesium, manganese, vitamins E, and B. In addition, nuts are generally a superb substitute for gluten.

3. Seeds

Seeds are the small factors of flowering vegetation that help in replica. They include proteins and nutrients, which include calcium, iron, and magnesium. Seeds can be eaten raw or roasted. However, continuously take into account that roasting the seeds receives rid in their dietary blessings. Seeds are an amazing supply of proteins, carbohydrates, and fiber.

four. Mushrooms

Mushrooms are the not unusual thickenings that expand on soil or wooden and include large proteins. According to current research, mushrooms are fantastic for health due to the reality they may lower or growth blood strain. They are wealthy in vitamins and minerals like zinc and potassium. In addition, they lower the hazard of coronary coronary coronary coronary heart disease and are notable for the immune gadget, blood sugar, and anxious device.

5. Vegetables

While most veggies are farmed, a few are wild veggies which you need to realize about. Wild veggies are wealthy in vitamins because of the fact they grow obviously without human interference. They have a better zinc, iron, and magnesium content fabric. Vegetables consist of some of fiber and carbohydrates vital for the proper functioning of the digestive device. Vegetables are also essential for the cardiovascular system and decrease blood pressure.

6. Herbs

Herbs are flowers used for flavoring, medicinal functions, and perfumes. Their materials are typically risky, aromatic compounds. Although a few herbs are toxic if now not properly organized, maximum are tremendous in your fitness. Some herbs embody tannins (catechin) and antioxidants, and some act as diuretics and expectorants, at the same time as others have an effect on the stomach simply.

7. Nettle

Nettles are common weeds normally discovered in wet areas. They incorporate calcium, phosphorus, potassium, magnesium, and nutrients C, are rich in fiber, and are often used as an alternative for spinach due to the fact they contain more iron. Nettles want to be eaten cautiously and steamed in advance than eating. When nettles are cooked, they lose their stings and come to be fit for human intake.

8. Dandelion

Dandelions comprise a excessive quantity of vitamins A, C, potassium, calcium, sulfur, iron, phosphorus, magnesium, manganese, and vitamins B1 and B2. This weed is used for one-of-a-type features, in food or medicine. When dandelions are eaten raw, they smooth the urinary tool and help with digestion. When eaten with lettuce, they help lessen levels of cholesterol in the blood, growth urination, and modify the belly. Dandelion tea permits with urge for food, at the same time as dandelion roots can assist prevent diarrhea.

9. Raspberry

Raspberries are commonly red or black and include many vitamins and minerals, along with calcium, magnesium, phosphorus, folate, iron, manganese, nutrition A and C, fiber, carbohydrates, and proteins. Raspberries are notable for the digestive machine due to the fact they've got fibers that clean the stomach walls thru casting off pollutants. They also are very rich in antioxidants and help save you coronary coronary heart infection.

10.　　Hazelnut

Hazelnuts are normally brownish-red or yellow-brown nuts that expand on bushes. They are wealthy in carbohydrates and minerals like calcium and iron. Hazelnuts additionally have an extraordinary quantity of protein and fiber. However, they have to now not be eaten in greater due to the fact they could purpose weight benefit.

11.　　Chestnut

Chestnuts are brown nuts that expand on timber. They are wealthy in carbohydrates, fiber, protein, and minerals like iron. Chestnuts are commonly cooked or boiled in advance than being ate up due to the reality they'll be hard to digest if eaten raw. Some chestnuts include immoderate quantities of tannins that purpose diarrhea, constipation, vomiting, stomach ache, gas, and cramps.

12. Acorn

Acorns are nuts that increase on timber. They contain carbohydrates, fiber, protein, and minerals like calcium, magnesium, phosphorus, zinc, and iron. Acorns are used to make flour for bread due to the fact they'll be wealthy in carbs and fiber. However, acorns should be boiled earlier than consuming because of the fact a few can cause diarrhea or vomiting at the same time as ingested uncooked.

thirteen. Sunflower Seeds

Sunflower seeds are used for food and decoration. They consist of many minerals, which includes calcium, iron, magnesium, manganese, phosphorus, and zinc. Sunflower seeds additionally have excessive quantities of protein and fiber that gain the cardiovascular machine. In addition, they'll be rich in vitamins E, which allows with imaginative and prescient and pores and skin health. Sunflower seeds moreover comprise antioxidants. However, sunflower seeds want to not be excessively eaten because of the fact they're capable of motive weight advantage, weight troubles, lower blood sugar tiers, and high levels of cholesterol.

14. Pumpkin Seeds

Pumpkin seeds are small inexperienced or yellow seeds that grow on orange pumpkins. They have masses of iron, calcium, magnesium, phosphorus, potassium, zinc, vitamins B1, and manganese. Pumpkin seeds are commonly eaten raw but additionally may be roasted. They encompass antioxidants that protect the frame from external pollutants. However, pumpkin seeds ought to no longer be excessively eaten because of the truth they're able to motive weight advantage and immoderate levels of cholesterol.

15. Wild Strawberries

Wild strawberries are small red plant life that growth on the floor. They encompass hundreds of fiber, diet B1 and C, potassium, magnesium, manganese, and calcium. These plant life can be eaten raw or used in brilliant recipes like salads or cakes (irrespective of the fact that they can not be used to make jam or jelly).

sixteen. Wild Blueberries

Wild blueberries are small blue flowers that develop on the floor. They are rich in fiber, weight loss plan C and K, potassium, magnesium, manganese, calcium, and iron. Wild blueberries can be eaten raw or used in one-of-a-type recipes like salads or cakes. It's notable to devour them cautiously due to the truth they may be able to purpose diarrhea or stomachache if eaten excessively. The leaves of wild blueberries are also steady to consume.

17. Wild Plums

Wild plums are small darkish purple plant life that develop on wooden. They include minerals, which encompass potassium, zinc selenium, calcium manganese, and magnesium. Wild plums may be eaten uncooked or utilized in recipes. It's first rate to consume them fairly because of the reality they may reason diarrhea, stomachache, gas, and bloating if eaten excessively. Plums additionally incorporate numerous herbal sugars and want to be eaten in small portions.

18. Wild Asparagus

Wild asparagus is a green plant that grows within the floor. It is excessive in fiber, nutrition C and K, potassium, magnesium, calcium, manganese, and iron. Wild asparagus is generally eaten uncooked in salads however additionally may be boiled before eaten. It's quality to devour wild asparagus reasonably due to the fact it may cause bloating, stomachache, nausea, and diarrhea if eaten excessively. Proper cooking is likewise very vital.

19. Watercress

Watercress is a green plant typically grown in the water. It contains lots of fiber, nutrients A and C, potassium, iron, calcium, manganese, and magnesium. Watercress is generally eaten uncooked in salads but also can be boiled earlier than eaten. When eaten carefully, watercress blessings the cardiovascular tool and allows vision fitness.

20. Black Currant

Black currants can be eaten glowing or made proper into a puree and utilized in distinct recipes. They are rich in food plan C, potassium, magnesium, manganese, calcium, and iron. Black currants may additionally moreover cause stomachache or diarrhea if eaten excessively. Therefore, humans with kidney stones should keep away from consuming black currants. It's precise to be conscious that black currants also can be used to make wine.

21. Blackberries

Blackberries are round and colourful dark purple fruits that increase on thorny wood. They comprise excessive vitamin C, omega-three fatty acids, potassium, magnesium, manganese, iron, and calcium. Blackberries are commonly eaten raw but may be carried out in special recipes like desserts or pies (despite the reality that they can not be used to make jam or jelly). Although they consist of many vitamins, blackberries have to be eaten reasonably because of the truth they are capable of cause diarrhea, stomachache, fuel, bloating, or nausea if eaten excessively.

22. Elderberries

Elderberries are darkish blue give up end result that expand in thorny wooden. They are wealthy in vitamins A, healthy dietweight-reduction plan C, omega-three fatty acids, potassium, magnesium, manganese, iron, and calcium. Elderberries are used to make wine, jelly, syrups, and compotes. Elderberries can be eaten uncooked fairly, but they must no longer be fed on excessively. The leaves and blossoms of elderberry vegetation additionally may be eaten raw.

23.   Sugar Maple Tree

Sugar maple wood are medium to large wooden that boom in North America. They include plenty of nutrition C, potassium, manganese, calcium, and magnesium. Sugar maple bushes are tapped in past due wintry climate or early spring to make maple syrup. The syrup is used for particular recipes like pancakes, waffles, sweets, and fudge.

24.   Wild Chicory

Wild chicory is a inexperienced plant that looks like a daisy. It consists of weight loss plan A, calcium, iron, and magnesium. The leaves of chicory plant life can be eaten uncooked or boiled in advance than being eaten. When eaten cautiously, chicory leaves advantage the cardiovascular tool. The chicory root additionally may be eaten and is normally cooked or roasted in advance than consumed.

25.   Wild Mustard

Wild mustards are inexperienced vegetation with small yellow flora that develop inside the floor. It consists of omega-3 fatty acids, calcium, magnesium, thiamin, potassium, healthy dietweight-reduction plan C, and manganese. The leaves of wild mustard may be eaten uncooked or boiled earlier than being eaten. After boiling, the leaves also may be implemented specifically recipes like salads, sandwiches, and soups.

26. Wood Sorrel

Wood sorrel is a inexperienced plant with 3 first-rate leaves on every stem. They comprise nutrients C, calcium, potassium, and magnesium. The leaves are very bitter, so they are generally eaten raw in salads. The plant life of wood sorrel vegetation also are fit for human intake and can be eaten uncooked in salads.

27. Wild Garlic

Wild garlic is a inexperienced plant that has small white plant life. It has a strong scent, so it's far utilized in cooking in area of raw. Wild garlic is rich in food regimen C, calcium, iron, magnesium, manganese, and potassium. The leaves of this plant may be eaten raw or boiled. After boiling, the leaves may be carried out in wonderful recipes like soups, salads, or seasoning.

28. Plantain (Plantago)

Plantain is a inexperienced plant with long leaves. It incorporates omega-3 fatty acids, calcium, magnesium, potassium, vitamins C, and manganese. The leaves of plantain flowers may be eaten raw or boiled in advance than being eaten. They moreover may be implemented in special recipes like salads, soups, or as a seasoning.

29. Mint (Mentha)

Mints are inexperienced plant life with small pink vegetation. They include immoderate omega-3 fatty acids, calcium, magnesium, potassium, nutrients C, and manganese. Mint leaves may be eaten uncooked in salads or on sandwiches, however they should be eaten moderately because of the reality they may be capable of reason flatulence. Mint leaves are utilized in distinct recipes like jellies, syrups, beverages, and sweet.

30. Mallow (Malva)

Mallows are inexperienced flora with small flowers. They comprise calcium, magnesium, potassium, and nutrients C. The leaves of mallows can be eaten uncooked or boiled in advance than being eaten. They are applied in particular recipes like salads, soups, or as a seasoning.

31.   Thistle (Cirsium)

Thistles are inexperienced vegetation that expand thorns. The leaves and stems can every be eaten cooked or raw in salads. They additionally may be used for remarkable recipes like soups and salads. Thistle roots can be eaten after being boiled or roasted. The roots additionally can be used for taken into consideration one in every of a kind recipes like fritters or whilst a espresso opportunity.

32.   Purslane (Portulaca Oleracea)

Purslanes are inexperienced plants with small yellow flora. The leaves of purslane vegetation may be eaten raw in salads or sandwiches. What's particular approximately purslane is that it can additionally be eaten cooked like spinach. Purslane leaves encompass omega-three fatty acids, calcium, magnesium, potassium, vitamin C, and manganese.

33. Buckwheat (Fagopyrum Esculentum)

Buckwheat is a inexperienced plant that has white or pink plant life. The leaves, flora, and seeds of this plant are all steady to eat. The leaves may be eaten uncooked in salads or as a seasoning, but they must best be eaten pretty because of the truth they consist of oxalates. It's first rate to boil or roast the leaves earlier than eating them. The flowers also may be eaten raw in salads, and they encompass masses of nutrition C, calcium, and magnesium. The seeds can be boiled, roasted, or become flour and utilized specially recipes like pancakes or quantities of bread.

34. Violets (Viola)

Violets are inexperienced flora which have plants with 5 petals. The leaves of this plant are appropriate for consuming, however they want to by no means be eaten raw because they comprise plenty of oxalates. The leaves can be eaten boiled or applied in a single-of-a-kind recipes like salads and omelets, jellies, syrups, and beverages. Violets leaves comprise masses of calcium, magnesium, potassium, and food regimen C. So, if you need to feature calcium and magnesium in your eating regimen, the leaves of violets can be a superb way to do it.

35. Sharp-Leaved Plantain (Plantago Lanceolata)

Plantain is a green plant with lengthy leaves. Usually, the leaves of plantain plants are massive and flat, but sharp-leaved plantain leaves are lengthy and narrow. The leaves of this plant can be eaten raw or boiled earlier than being eaten. After boiling, the leaves can be used for precise recipes like soups, salads, and even seasoning.

36. Chicory (Cichorium)

Chicory is a green plant with small blue or lilac plants. It includes lots of calcium, magnesium, potassium, and vitamins C. The leaves of chicory plant life can be eaten raw in salads. They additionally may be used for exceptional recipes like soups and salads. Chicory roots may be eaten after being boiled or roasted. The roots also may be used for exquisite recipes like fritters or even as a coffee opportunity.

37.   Curly Dock (Rumex Crispus)

Curly dock is a inexperienced plant that has lengthy leaves with wavy edges. This plant includes hundreds of magnesium, potassium, and nutrition C. The leaves of curly dock flora can be eaten raw in salads or as a seasoning. After being boiled or roasted, they can also be eaten in particular recipes like soups and salads, omelets, or on the identical time as a espresso opportunity.

38.   Burdock (Arctium)

Burdock is a inexperienced plant with purple flora and thorns. The suitable for ingesting elements of burdock plants include their leaves, roots, and seeds. The leaves of burdock vegetation can be eaten uncooked in salads or as a seasoning. Roasting, boiling, and frying the leaves are also brilliant methods to put together them in advance than consuming. If you want to feature some potassium, calcium, and magnesium into your diet regime, then burdock leaves may be a splendid way of doing that.

39.   Pineapple Weed (Matricaria Discoidea)

Pineapple weed is a green plant with yellow flowers. The leaves of pineapple weed plant life may be eaten raw in salads or as a seasoning. After being boiled, in addition they may be eaten in special recipes like syrups or maybe seasoning for stews and salads. Pineapple weed leaves encompass masses of diet plan C, calcium, and magnesium.

forty.   Ground Ivy (Glechoma Hederacea)

Ground ivy is a inexperienced plant with small pink plants. The leaves, stems, and plants of ground ivy flora can all be eaten. After boiling or steaming the plant components, they may be eaten as a aspect dish or used for precise recipes like omelets and salads. Ground ivy consists of hundreds of vitamins C, magnesium, potassium, and calcium.

The Differences among Plants, Flowers, Weeds

There are many variations between plant life, plant life, and weeds. The important differences encompass wherein they develop, their use, and the manner they appearance. While maximum plants are placed into the magnificence of "plants," there may be a subcategory referred to as weeds. Weeds are considered any plant deemed as bothersome, unwanted, or complicated.

Flowers are residing organisms with colourful components to draw insects and exceptional animals. Some vegetation are taken into consideration weeds, however not all. Weeds expand everywhere and spread hastily sooner or later of the environment. On the opportunity hand, vegetation generally develop in unique locations and unfold slowly.

Here's a comparison table of plants, flora, and weeds:

Plant Flower Weed

Has roots Is the reproductive a part of the plant Doesn't have any roots

Needs daylight hours, water, and nutrients to develop Needs sunlight hours, water, and nutrients to develop Grows rapid, taking vitamins and water from great vegetation to stay to inform the tale

Produces plants or seeds that grow to be new vegetation Only lasts for a quick time

Produces seeds dispersed via using birds, wind, or water

Plants are dwelling organisms with roots, stems, leaves, plants, and seeds. Plants want daylight hours, water, and nutrients to increase. When they mature, flowers produce flora or seeds that in the end become new flowers. Plants are normally divided into categories, such as timber, shrubs, vines, grasses, ferns, mosses, algae, and flowers.

Flowers are the reproductive organs of vegetation. They are normally colorful with petals that entice pollinators. Flowers simplest final for a fast time after which die. Flowers are usually divided into 4 commands: wildflowers, lawn plant life, ornamental plants, and weeds.

Weeds are vegetation that don't have any roots and increase quick, taking nutrients and water from distinctive plant life to live on. They reproduce thru the usage of dispersing seeds carried thru birds, wind, or water. Weeds are living organisms and increase in locations like the cracks of sidewalks, grass, and soil. They grow without humans being concerned for them and commonly unfold rapid.

Edible wild flowers provide natural alternatives to business meals items. They are used for emergencies and regular, each day existence. Edible wild plants were part of the human weight loss plan for lots of years. They provide a dependable supply of carbohydrates, proteins, micronutrients, fiber, and wonderful vitamins to preserve the body healthful.

Like every different meals, suit to be eaten wild flowers may be poisonous if now not used properly. Some in form for human intake wild flora are poisonous while eaten raw however safe to devour even as cooked. Others are handiest poisonous in excessive high-quality seasons or stages of increase. A toxic plant may additionally look nearly identical to an stable to consume wild plant, so it is crucial to find out every plant efficaciously before ingesting it.

Chapter 2: Are Wild Edible Plants Safe To Eat?

People have used wild in form to be eaten vegetation for hundreds of years. Today, many human beings are inquisitive about consuming extra natural, neighborhood, and natural food to lessen their carbon footprint and enhance their fitness. Many human beings ask if it is secure to eat wild steady to eat vegetation. Some poisonous plant life increase in the wild, but maximum wild healthy for human consumption vegetation are safe to devour and are observed nearly everywhere. Many wild suit for human consumption flora are pretty huge, but some are uncommon or perhaps endangered. It is important to recognize the plant to eat it thoroughly as some wild wholesome to be eaten plant life can cause detrimental reactions in human beings. These flora are not stable to consume until they may be nicely recognized. Many wild match for human intake plants taste perfect and are used within the identical way as common grocery keep produce.

Most wild wholesome to be eaten plants may be eaten raw or cooked. Some should be very well wiped clean of grit and bugs first. Others, which includes acorn, are nutritionally stepped forward with the useful resource of way of lightly roasting or baking. Plants may additionally additionally want to be strained for sand and grit, parboiled in severa adjustments of water, or leached with severa modifications of boiling water. Leaching is crushing or pounding the plant meals, placing it in a strainer, and pouring boiling water via it to remove unstable substances.

Be sure of the plant's identity in advance than eating it, and in no way consume wild suitable for eating plant life you aren't familiar with. Many wild appropriate for eating plants appearance similar to toxic plants. Some wild mushrooms are deadly poisonous regardless of the truth that specific, intently associated wild mushrooms are secure to eat. This bankruptcy gives fundamental facts about the advantages of consuming wild suitable for ingesting flora and nicely identifying common ones. It'll assist you perceive which additives are in shape for human consumption and which aren't.

Benefits of Eating Wild Edible Plants

The blessings of eating wild healthy for human intake plants are many. Many humans prefer to eat local food that's clean, freed from chemicals and insecticides, nutritious, and scrumptious. Wild in shape for human intake plants are simply that. They offer healthy, glowing meals with the minimum environmental impact. Foraging wild suit for human intake vegetation moreover can be a a laugh and profitable interest similarly to the precept benefits of real vitamins and terrific flavors. It is a manner to connect to nature and experience workout outside.

• They are licensed herbal.

• The plant is hardy and grows without problem. It's proper for the environment and doesn't need to be transported over prolonged distances.

• Most wild healthy to be eaten flowers are pretty huge. You can forage right now in your backyard or neighborhood park.

• In generally, wild healthful for human consumption vegetation are eaten uncooked or used in recipes.

- A big type of match to be eaten plants are to be had. You may additionally even invent your recipes.

- Most wild healthful for human intake flora aren't expensive.

- They are free of herbicides, insecticides, and one-of-a-kind chemical substances located in commercial employer food.

- You can experience happy with eating nearby meals this is glaringly grown.

Edible flowers provide many blessings to the human frame. They are low in energy and excessive in nutrients, offer fiber and carbohydrates, and include sufficient quantities of zinc, iron, and calcium. Many wild in shape to be eaten plant life help with digestion or are a slight laxative. Some wild fit to be eaten flora are used as a herbal remedy for loads not unusual ailments, collectively with complications or disillusioned stomachs.

Identify the Edible Parts

It's important to recognise a few protection basics to understand which parts of the plant are in shape to be eaten. It is critical no longer to consume any a part of a poisonous or unknown plant. A poisonous plant can be deadly even within the smallest quantity. Some plants have safe to devour and toxic components, so that they want to be identified cautiously. It's best only to devour plants you could in truth select out as steady. There are many strategies to discover a plant without consuming it. Some traditional methods to apprehend a plant are:

1. Examine the Leaves

The leaves are the most beneficial part of the plant for identity. The leaf shapes, patterns, margins (edges), and veins help to come to be aware about a plant. Most wild safe to eat flowers have either jagged or smooth leaf margins. If the leaf has no stalk (petiole) and is set up without delay to the stem, the complete plant may be appropriate for ingesting. The veins want to no longer contain white filaments. If the veins are whitish or "bushy," do not consume that a part of the plant. It's critical to ensure the leaves don't have spines as they'll worsen your throat. If you aren't tremendous about a plant, do NOT consume any a part of it.

2. Examine the Stem

The stem is some other crucial part of identifying a plant because of the fact the leaves are connected to it. A leaf can assist decide if it's fit for human consumption. The stems of maximum wild stable to consume vegetation are each fleshy or woody. A fleshy stem is softer and can be eaten like a vegetable. It is vital to differentiate a fleshy stem from a toxic or unknown plant. Make high quality the stem isn't hollow and includes no milky sap. A woody stem is used to make tea or one in all a kind drinks, but it should be peeled. Woody stems are typically not advocated for ingesting due to the reality they are too tough and fibrous.

three. Examine the Flowers

The flowers help find out a plant, and maximum wild in shape for human consumption plants do no longer have placing flora but are determined in lots of specific kinds. If a plant has flashy flowers, it is not always poisonous. Some people may be allergic to pollen from plant life, and it's going to cause rashes. The plants need to no longer have huge petals that spread out like a movie star shape due to the truth they will be poisonous. If the flower is small, has many petals, and spreads across the duration of the stem, it is probably in form to be eaten. Flowers typically taste candy or slightly sour. The fragrance is also a extremely good indicator of whether or not or no longer or not a plant is safe to devour. A flower with a candy odor is maximum possibly steady to devour, and plant life with a awful fragrance are in all likelihood toxic.

4. Examine the Seeds

The seeds can assist to understand a plant in some instances. If the seed is big, round, and smooth to choose out as a fruit or nut, it's far suit to be eaten. However, no longer all seeds look this way. Some flowers have suitable for ingesting seeds, and others have inedible or poisonous seeds. The appearance of a seed can be deceptive because it varies from plant to plant. It is first rate to keep away from consuming seeds that have no longer been identified as secure or risky. Many seeds are too small and want to be scraped off the seed head together together along with your teeth. Some wild in shape for human intake flowers have seeds that appear to be brown or black dots or fibers. The brown fibers may be scraped off and eaten, however the black dots have to now not be eaten.

5. Examine the Root

The plant's root may be used to choose out it and provide key facts about whether or now not or now not or now not a plant is match to be eaten. The root ought to be white indoors because of the truth the shade yellow may moreover moreover suggest poison. If the foundation is grayish or brown, it could be secure to eat if it isn't always sour. The color of some vegetation modifications at the same time as the plant is cooked, making them secure to consume. If you are unsure whether or no longer or no longer or no longer the premise is poisonous, do now not eat it.

Make effective to investigate in advance than eating a wild suitable for eating plant for the number one time. It is critical to ensure that the plant isn't always poisonous due to the reality ingesting a poisonous plant may be lethal. Always use caution even as selecting to eat a wild in shape to be eaten plant.

Once you have got decided if the plant isn't always poisonous, you could devour it. Never eat wild plants raw or raw till you're certain they aren't toxic. Some plant life lose their pollution as fast as processed but are toxic of their raw form. The poisonous factors of the plant are lots extra focused inside the uncooked form, so as quickly as a plant has been cooked, the poisonous additives disappear, and it becomes strong to consume. If that a excellent part of a wild safe to eat plant need to no longer be eaten due to the reality it's miles toxic, dry it out and do away with it.

Ways to Ensure the Plant Is Safe to Eat

Field publications are useful on the same time as figuring out flora you are not acquainted with. Many display photos of the plants and provide certainly one of a kind records approximately them. This may be very useful at the identical time as you're unsure if the plant is toxic or suitable for eating. If you're new to foraging for food, it's miles quality to purchase a guide and use the images and statistics supplied.

Never eat some issue no longer diagnosed as a consistent plant or meals. If you cannot apprehend the plant, do not eat it until you have determined what it's far. It is important continuously to be careful whilst foraging because you do no longer understand what plant life or elements of plant life are toxic and in no way devour wild in shape to be eaten plant life besides you're nicely knowledgeable approximately them.

If the plant is listed as appropriate for eating however has a bad scent, search for brilliant plant life with out a lousy scent. If the plant isn't always listed in a guide, avoid consuming it because it may be toxic. Every manual may additionally additionally listing the same plant as in form for human consumption, but one-of-a-kind courses will list particular elements of the plant as suit for human intake. It's moreover critical to apprehend the suitable methods to eat and prepare the ones plant life. Here's a list of the most commonplace strategies to prepare wild secure to eat flora.

Preparation of Wild Edible Plants

1. Rinse

Rinse the plant underwater to put off dirt and any insects in advance than eating it. The insects may be toxic, and consuming them will make you sick. It is essential to make sure the plant life you forage for food are secure and free of bugs. If you discover a pc virus to your plant, remove it carefully with out squishing it. Check over the plant very well to make sure no insects are hiding within the folds or crevices earlier than ingesting it. Also, take a look at the leaves of plants for aphids because they may be generally placed on them.

2. Boil

Boil the plant for as a minimum 5 minutes to put off any poison within the plant. Many toxic plant life are not dangerous if eaten after they have been boiled, however it's far but an terrific idea to boil them, absolutely in case. If the plant has been boiled, it is stable to consume, and you may devour any part of the plant. After boiling the plant, check if it has a awful or sour taste. If it does, then it is not safe to eat.

3. Add to a Meal

Just due to the truth a plant can be eaten uncooked does now not mean it must be eaten uncooked. Some plants are extremely good at the same time as delivered to a cooked meal in preference to eaten raw.

Make best to prepare dinner the plant earlier than ingesting it in case you aren't nice whether or not or no longer it is able to be eaten raw because of the reality many flowers can be toxic while eaten raw.

four. Dry

You can dry vegetation in the sun or in a dehydrator to preserve them for later use. Dried flora can be stored in a subject till they may be organized for use. You should make soup with dried plants, located powdered dried flora into smoothies, or devour them as a snack. If you dehydrate the plant, make certain to boil it earlier than ingesting. During wintry climate, you may also make tea with dried flowers like rose hips and hibiscus flowers. When you boil dried plant life, there can be no want to peel them. You can without difficulty dry your plants or buy dried flowers from a shop.

5. Juice It

Juicing is a fun manner to eat wild suit for human intake flora. Some additives of a plant, including roots and flowers, cannot be eaten uncooked because of the reality they will be toxic. You can juice them if you boil them first. Preparing the roots and plant life on this manner makes them safe to consume.

6. Make a Tincture

Tinctures are a extremely good manner to get the advantages of vegetation with out ingesting them. Many flowers can be dangerous at the same time as eaten, however they'll be regular on the equal time as ate up as a tincture. Research instructions on-line to make your tinctures if you do not want to buy one from a shop.

Adverse Reactions

It is important to don't forget that detrimental reactions are not not unusual on the same time as ingesting wild flowers. If you experience an destructive reaction, it's far high-quality to touch a medical doctor. If you positioned the plant is poisonous, do not consume it or take the plant's components with you to your medical health practitioner. Here are some subjects to appearance out for:

1. Stinging, burning, or itchy pores and pores and skin

2. A rash or hives

three. Organ damage (kidney, liver)

four. Blood inside the urine or stool

5. Seizures or respiratory problems

6. Fever or chills

7. Nausea, vomiting, upset stomach

eight. Diarrhea

9. Headache

10. Unconsciousness

11. Difficulty strolling

12. Disorientation

13. Some flowers also can motive beginning defects in pregnant girls. If you're pregnant, do now not devour wild plants till a health practitioner has authorized them.

With research and a chunk of exercise, people can learn how to discover, prepare, and enjoy suitable for eating wild flora appropriately.

General Tips

1. Learn about the plant you are eating. If you do no longer understand what the plant is, it is great to keep away from ingesting it. Many

steady to consume plants seem like poisonous plants. If you aren't notable approximately the plant, take pictures of the complete plant and its additives to observe later.

2. Carry a ebook about healthy to be eaten plants with you at the same time as you skip in your foraging expeditions. This is in particular important in case you are by myself because of the reality in case you ingest a toxic plant, it'll assist the individual that reveals recognize which plant delivered on the dangerous response.

3. Practice makes perfect. Before starting, exercise identifying and consuming those plants with an skilled forager or a guide.

four. Wild flora want to be washed before ingesting them, even though they'll be boiled or cooked. If you do not wash the plant first, it would make you unwell. Do this for all factors of big plant life and roots as a manner to no longer be peeled in advance than cooking or eaten raw due to the fact their outer layer includes pollutants that have to be washed off.

5. If you go on a long hike or foraging day adventure, % a lunch in preference to searching out healthy to be eaten plants along the way. You can acquire in form to be eaten flowers along the direction after ingesting lunch and have greater power from being well fed in advance than starting your journey.

6. Be careful at the identical time as ingesting unknown flowers. Even when you have prepared it efficaciously, you could have a lousy response to the plant.

7. If a person is with you, take turns consuming one small a part of the plant first to look how your body reacts.

eight. Some of those wild safe to consume plant life may be toxic, so do now not consume them raw or devour an excessive amount of right away.

9. Keep a number one useful beneficial aid kit with you actually in case someone has an unfavourable response to the plant.

10.   Do now not devour something until whilst you check at the side of your medical doctor to make sure it's far secure.

Edible flora may be eaten and offer the body with dietary rate in addition to sugars, starches, and oils. Wild flowers and flowers long past to seed are also appropriate for eating. Many wild suitable for eating plant life are found in a few unspecified time in the future of the area, so are looking for advice from your nearby library or internet for a list of untamed suitable for consuming vegetation.

Wild in shape for human consumption plant life are an extremely good food deliver and remedy, however some components of the plant are more nutritionally useful than others. Eat the plant life, roots, leaves, stems, or seeds to get the most out of the plant.

Edible flowers are stable to devour if organized correctly, so consuming the right elements of the plant is useful in your fitness. It is important to bear in mind that a few wild secure to devour flora are poisonous. If you do no longer understand what the plant is, it's far incredible to avoid it. Do no longer devour any vegetation unless you're knowledgeable sufficient to recognise they're steady to consume and popular with the beneficial resource of a health practitioner.

While foraging, take photographs of the whole plant and factors so that you can examine them later. Learning about the vegetation earlier than ingesting them is essential. Practice identifying and ingesting wild vegetation with an skilled forager and maintain a first resource bundle accessible. It is crucial to be careful while consuming unknown flora thinking about that you may have an bad response. Only eat one small part of the plant within the starting if a person is with you to check your frame's reaction. Wild in shape to be eaten plant life can be used as meals and treatment, however eating the right additives guarantees you get the finest advantages.

Chapter 3: Harvesting Edible Wild Plants

Before you start, you need to analyze a few fundamentals about wild plant life. The more information you advantage, the better your adventure of foraging wild safe to consume vegetation. When and what are you able to devour? Which components ought to you operate, and the manner to put together them efficiently? For instance, even as a few berries are sweet and attractive raw, others are higher at the same time as cooked together with meat.

Gathering them may be masses less difficult in case you recognize the plants you are seeking out. Make high best to print out an photograph of the plant you searching for because it makes identification an awful lot less complex at some stage in the adventure. It is usually higher to be stable than sorry whilst foraging. This monetary catastrophe carries an define of a few trendy regulations and hints on finding wild secure to devour plant life, easy foraging regulations, tools you ought to deliver, and in which to look.

Foraging Rules for Beginners

Before you even go out to look for wild wholesome to be eaten plants, you want to understand some simple pointers and information to preserve your journey secure and a success. Going foraging unprepared can cause intense stomach aches or possibly loss of existence.

Rule 1: Don't consume the whole lot you note

It is proper to be careful. If a plant seems delicious, there may be a reason for that. Only eat vegetation you comprehend and might come to be aware about with reality. If you are not 100% positive what you are ingesting, do no longer consume it. The equal is going for internal organs. Only devour the elements of plants you're excessive excellent approximately. It's brilliant only to devour the fruit to begin with and slowly add superb components of a plant in your eating regimen. For example, first great eat the fruit, later the leaves, after which the roots.

Rule 2: Never take extra than 50% of a plant.

More frequently than no longer, wild safe to devour flowers broaden in great numbers. Don't take too much of a plant; leave enough for exclusive animals and for the plant to reproduce. It is in no way smart to take extra than you want. When you forage wild appropriate for eating vegetation, continuously leave some of the fruit, leaves, or other elements in the back of due to the fact the plant will live on, and you're ensuring it does now not disappear from your environment.

Rule three: Be aware about seasonality.

It is brilliant only to consume wild appropriate for consuming plants when they are in season, or as a minimum realise at the same time as they're in season. Most flowers aren't to be had all 12 months round, that means you need to apprehend the right season for harvesting. For instance, berries are most effective to be had in some unspecified time in the future of the summer season months. Some vegetation like cattails or water lilies can be found all one year round in case you appearance within the proper areas.

Rule four: Be privy to your surroundings.

During foraging, you should be aware about your environment. Of course, you don't have to be paranoid about it, however you need to generally preserve one eye open. Be alert of animals in the place and be cautious for toxic flora. Be careful of your pathway, too, so you don't harm the surroundings unnecessarily. When foraging wild in shape to be eaten plant life, it's first-rate to stay on trails you've created. If there are none, make certain you are nicely aware about in which you're going and a manner to get out.

Rule five: Always tell a person in that you are going.

The final rule to foraging wild match for human consumption flowers is constantly to allow a person recognize wherein you're going and even as you may pass decrease again. If you do now not go back on the agreed time, a person would possibly in all likelihood end up worried and start searching out you. This ensures your protection want to you wander off or come across some other limitations.

Basic Rules for Foraging

You want to usually have a sure technique to conform with while you are foraging wild suit for human intake vegetation. It ensures the protection of you and your surroundings. These crucial hints may be tailor-made relying on whether or no longer or now not you are with someone.

1. Bring the Right Tools

Before you go out foraging, it's far fantastic to build up your equipment first. It is critical to have robust baskets or baggage available to hold your findings. Make certain you have a field to boil water in if desired and produce some packing containers on your harvested flora. To hold yourself solid, supply a knife with you. You can use it to reduce toxic flowers or vines that is probably blocking your way.

2. Always Look in advance than You Step

When foraging wild appropriate for eating flowers, usually searching in which you are going is crucial. Watch out for poisonous vegetation and in no way step on a mushroom. There might be a nest hidden underneath the plant you step on, and you don't want to disturb any pink ants. Always appearance in which you're going, and you will keep away from bumping into matters.

three. The Right Materials

When foraging, it's far essential to deliver the proper materials. You want to supply a box to vicinity the flowers in. Make sure this container is straightforward and has no holes as it needs to be watertight. It's first-rate to apply baskets or plastic baggage. Pick baskets also are beneficial if you are simplest selecting cease result. You might want to apply your fingers at times to pick out out up touchy berries, so make sure they are easy.

4. Time of the Day

Going foraging inside the morning is first-rate due to the fact the flowers are damp with dew and no longer withered from the solar. This is also the nice time to find out cease result because of the truth they may be softer and juicier within the morning, in particular if they're ripe. You can locate flora a great deal less complicated in the morning, too. Look for closed ones and open them to test if they're steady to devour. Just be careful now not to the touch any poisonous flora, as they'll switch their pollution to your frame.

five. Safety First

Always maintain protection first whilst foraging. It is notable to move foraging with anyone, especially in case you are new to it. You also can take an expert character for the number one time and look at from them. It's fantastic to use a guidebook, so always preserve one handy to make sure you apprehend what you're selecting.

6. The Direction of the Sun

To tell which flora are which, you want to realize wherein they extend. The ones that grow towards the east are often more sour. Those within the south are frequently sweeter. So, it's satisfactory to begin within the north first and pick your manner closer to the south. It is important to realise that the extra bitter the plant, the healthier it usually is for you.

7. Check the Seasons

It is wonderful to select the right seasons, which range counting on in which you stay. Some in shape to be eaten plants can handiest be picked inside the path of a certain time of the year. Look up whilst those are for your place or ask an skilled character. Some flora genuinely have particular tiers of growth, so it's miles crucial to look for the right time. The great manner to recognize while to pick the in shape for human consumption wild vegetation is to apply a guidebook.

eight. Know the Plant You Are Looking For

When foraging wild suitable for ingesting flora, you want to recognize the exact plant you're searching out. It is high-quality to choose the commonplace ones native on your area because the ones usually develop in abundance. Make excellent you apprehend their names and the way to tell them aside from poisonous plants. Learn the appearance of the plant right all the way down to its leaves, bark, flower, fruit, and roots. If you need to be more secure, it's fine to have photos of them as a manual.

9. Beware of the Edible Ones Too

It is vital to apprehend that some vegetation might be suitable for consuming, however they will be toxic if they may be no longer organized in the right manner, like boiling. This typically takes region even as you pick out your materials in advance than they're ripe. Only pick ripe end result, leaves, and roots in case you need to ensure they're consistent to devour.

10. The Right Harvesting Method

When you want to acquire stop end result, it is terrific to acquire them via the use of hand. If they may be on immoderate branches, you may shake the wooden lightly to make the fruits fall. Just be cautious no longer to shake it too much due to the fact this can harm the fruit. If it's far a sensitive fruit, you could need to use your arms. Always make sure your hands are clean. You will want to dig roots up and pull them out of the ground. Some roots need to be cooked first before they may be eaten, and moreover they need to be peeled earlier than you consume them.

eleven. Don't Harvest If You See Pests on the Plant

It is crucial to realise that most appropriate for ingesting vegetation are not particular if they have pests on them, probably because of the reality those pests are in the soil wherein it is planted. If you need to be consistent, it's superb to move away the ones vegetation by myself. The maximum secure factor is to find out each other region.

12. What to Do When You Get Home

When foraging secure to consume plant life, what you do with them whilst you get home is similarly critical. The first element is to region the flora in water. This is to ensure they are but glowing and haven't wilted. The first-class manner to store the plant for later use is with a plastic bag in a groovy vicinity a protracted manner from daylight hours. This way, the plant can live glowing for longer. It might also keep pests away. You want to moreover alternate the water each 4 to 5 days and make sure that the plant remains glowing.

13.   When in Doubt, Leave It Alone

It is pleasant to head away the plant on my own at the same time as you're in any doubt. Rather ask folks who comprehend. Read up on it on-line or buy a ebook about safe to devour wild plant life, and if the e-book is small and reachable, take it with you. You want to in no way devour something you don't apprehend the way to put together or in case you're uncertain it's been organized correctly. Lastly, address your instincts. If the plant doesn't look right, it probable isn't. It's nice to move for a few other plant as an alternative if you aren't certain approximately the primary one.

Tools for Wild Edible Plant Gathering

Some gear make the enjoy a good deal less complex, extra steady, and more powerful. Without proper devices, wild suit for human consumption plant gathering can be cumbersome. Here are the equipment you need for amassing wild edibles.

1. Basket

Whether it's a portable basket or one you strap on your waist, it's miles to be had in available at the same time as accumulating extra than simply one plant. Have a basket organized whilst you're harvesting roots and shoots. The basket lets in you to gather extra immediately, freeing up greater time. It moreover protects the roots of fragile plant life from being overwhelmed. To make sure your bag received't be heavy, restrict it to 1-2 kilos of untamed in shape for human intake plant life. It's super to gather in smaller portions, mainly while the plant stays unknown. When it's time to transport domestic, truely vicinity your basket in a cool and darkish location.

2. Knife

Sharp knives are vital for wild in shape to be eaten plant accumulating. Whether you use a Swiss Army Knife or a entire-sized searching knife, the blade must be sharp and properly maintained. A silly knife will damage you while harvesting wild edibles, not to mention it'll take longer to reap every plant. You can get the project completed faster and more stable with a pointy knife. Be certain to keep your knife dry and clean. When no longer in use, shop it in a sheath or a knife roll. You can also preserve it in a plastic bag internal your pocket, presenting easy get right of access to.

3. Gloves

Leather gloves are best for harvesting roots. If you're into looking or fishing, a couple of leather-based-primarily based gloves is a in fact excellent wild in shape for human intake plant collecting companion. Leather gloves protect your fingers from thorns and additionally shield the muse. When digging up wild edibles, gloves hold your hands easy and solid from getting scratched through branches or burs. Wearing gloves even as gathering roots is vital due to the reality you'll get extra of the concept if it's no longer damaged. For shoots, it's wonderful to transport with out gloves.

four. Plastic Bags or Zip-Lock Freezer Bags

Freezing roots and culmination is an effective manner to maintain flora glowing for later use. You can hold them sparkling in plastic baggage or freezer luggage, however make certain to dispose of the air in advance than sealing the bag near. To make this much less hard, use a straw to suck the air out. Zip-lock freezer bags are also extremely good for preserving shoots glowing in case you don't plan on eating them right away.

five. Backpack

Backpacks are a ought to-have even as wild healthy to be eaten plant collecting. It's first-rate to place on it in the front instead of at the once more at the identical time as harvesting roots. A backpack continues your arms loose at the identical time as gathering wild safe to consume flora and helps you to deliver more subjects, because of this you'll have extra time to harvest. It moreover lets in you to take extra plant life domestic with out making more than one trips. The high-quality duration of a backpack is 42 liters, but the maximum crucial problem is that it's snug to place on.

6. Shears

These place of information scissors are used for harvesting shoots and especially bamboo shoots. They're outstanding for collecting shoots too tall to reap or people with branches at the manner up and people tough-to-gain regions. Make sure they're strong sufficient to reduce the plant with one snip. You can use shears to attain quit result, too.

7. Rake

A massive garden rake is quality for wild appropriate for consuming plant amassing. With this device, you may harvest stop end result faster and less complicated. You also can use it for leaf gathering or gathering seeds. A rake is a bit more time-eating than specific gear, but it'll get the process finished. The outstanding rake to apply is the claw kind – it receives into difficult-to-achieve locations and has an prolonged manage that humans can hold.

8. Fine-Toothed Comb

This device is ideal for harvesting culmination with hairy surfaces, like blackberries and raspberries. The tooth inside the tool separate the fruit from the stem, allowing you to build up masses in a brief quantity of time. The fantastic-toothed comb doesn't harm the fruit, so there's no need to fear about sacrificing first-class for amount. It's a brilliant tool to have in case you're going out for an extended time.

nine. Pruners

These handheld tools are ideal for small jobs, like harvesting a unmarried fruit from a department. They're moderate-weight and smooth to hold, making them high-quality for harvesting culmination from wood. Pruners are also used for harvesting wild edibles that develop underground, like desserts. These equipment are flexible and might harvest roots of all sizes, inclusive of the tiniest ones.

10. Containers

Using containers, you can maintain time and harvest more wild healthy to be eaten plant life. It moreover protects fragile roots from harm or crushing. You can also use plastic baggage or baskets for this motive. Plastic resealable baggage are ideal due to the fact they may be moderate-weight, making them precise for transporting outside gadgets. If you use baskets, ensure that the holes are small sufficient to preserve onto roots but huge enough to permit dirt bypass via.

eleven.   Scoop

A scoop or a shovel is critical for wild appropriate for ingesting plant harvesting. This tool is used to get amongst roots and eliminate whole flowers from the floor. If you're harvesting bamboo shoots, this device is important. It's extensively applied for planting or transplanting shoots to other locations. You can use a shovel or a spade, but you could additionally use your fingers. The latter approach is probably very time-ingesting and hard, so it's not encouraged for maximum humans. You can use a cooking strainer or large ladle to collect fruit or nuts on the floor.

High-satisfactory harvesters avail the ones gear to make their artwork much less hard and quicker, plus they're willing to spend money on an splendid backpack, shears, a rake, a fine-toothed comb, pruners, and containers to make certain they have got the whole lot they need for the pastime. Harvesting suitable for consuming wild plants is important for preppers, survivalists, and everyone who lives off the grid, far from civilization to check. The equipment indexed above are a beneficial addition for your survival toolbox. They assist you to gather meals faster and much less complex, making your wild suit for human intake vegetation foraging a more inexperienced hobby.

How to Gather Fruits and Nuts

The harvesting techniques for give up end result and nuts are top notch. You can eat them raw or shop them for later use. If you want to shop them, you want to first allow the give up result or nuts dry. They'll final longer, and you may consume them any time.

When foraging for nuts, ensure that they may be shelled and de-shelled effects by way of hand. It approach they're ripe and prepared to be harvested. If you find unopened nuts, shake the tree until the nuts fall off. Wrap them in a material bag, paper towel, or tissue to ensure they live dry. Store them in a groovy area some distance from daylight hours.

To harvest fruits, shake the wooden gently. The ripe fruit will fall off, and you may gather them with out hassle on the floor. Place them in a subject or bag just so they don't get dirty. If you word any berries with larvae or bugs, brush them off along with your arms. Fruits are suitable for eating once they're nonetheless organization, so don't allow them to over-ripen.

Edible wild plant life are an crucial meals source, however they're additionally vital to the environment. With right foraging strategies, you can hold a stability while imparting yourself with clean meals to eat. Harvesting in form to be eaten wild plants is a fun interest, specifically in case you go out into the wild without a distractions. It's additionally an great manner to exercising your frame. You can become an expert at figuring out steady to eat wild flowers with a piece exercising. Some of those vegetation may additionally appearance similar to toxic ones, however maximum are smooth to select out as speedy as you understand what to look for. Start with the not unusual suit for human consumption wild vegetation and regularly build your repertoire as you benefit greater revel in.

Chapter 4: Being Careful When You Forage

It's anticipated that over one-0.33 of the area's populace uses wild plants as a normal a part of their food regimen. But foraging may be dangerous besides you examine which species are appropriate for consuming and the manner to pick out them well. There are severa processes to understand if a plant is suitable for eating. Unfortunately, nicely figuring out a toxic plant can be definitely as difficult and probable greater risky. To differentiate a steady plant from its toxic look-alikes, you need to comprehend its traits. Learning the ones is time-consuming and includes large studies. The satisfactory approach is to analyze from human beings who have been the usage of wild plant life for meals for decades and train your youngsters the identical.

Children can learn how to recognize in form for human consumption or medicinal flora from a younger age. They moreover find out about the risks of poisonous flora and not to get them through accident. For the most difficulty, many toxic flowers are blanketed through vibrant shades or an ugly taste. But some toxic wild flowers don't put it on the market themselves thoroughly. It is tremendous to keep away from all wild flora till you could undoubtedly understand them as fit to be eaten or medicinal. Most poisoning is not deadly, but it's far though a excellent idea to keep away from ingesting them if the least bit feasible.

If you observed which you or a loved one has ingested a toxic plant and is probably life-threatening, are in search of for immediate clinical hobby. If it isn't lifestyles-threatening and you've were given the plant, take it in your doctor as a manner to identify it and cope with you for this reason. If you can't get in contact with a health practitioner at once, make a examine of your signs and signs and take snap shots of the plant, and whilst you come domestic, see a clinical health practitioner soonest.

This financial disaster offers a top degree view of handling toxic vegetation, data a number of the maximum risky or deadly toxic vegetation, and the manner to avoid them. An assessment of the maximum commonplace toxic plants in North America and some tremendous rules on well identifying stable to consume and medicinal wild flora is included.

The Common Signs of Poisonous Wild Plants

Many poisonous wild flowers are smooth to perceive because of their unpleasant taste. Generally, they've got an unappetizing color or scent. Poisonous flowers regularly incorporate chemicals we accomplice with the matters we dislike, which include natural alcohol, nasty chemical substances, or human waste. If a plant has this form of and you may't recognize it, do now not eat it.

Some toxic flowers resemble secure to eat plants so carefully that it's miles almost not possible to inform them apart. For those flora, you need to realize the overall trends of every plant family, which may be observed from a e-book or on line deliver. If you aren't up to speed on which plants are match for human intake, it's miles brilliant to live far from wild plant life until a person skilled can undoubtedly pick out them for you.

In maximum times, toxic flowers are smooth to discover and tough to confuse with unique species. If a plant has some of the following tendencies, it is also poisonous:

- Spines or first rate hairs.

- Parts of the plant are quite sour or in any other case unpleasant in flavor.

- Bitter or soapy flavor.

- Sap at the way to turn black at the same time as exposed to air.

- Leaves or buds which can be purple, crimson, pink, black, or orange.

- White berries.

- Seed pods which can be brown, grey, black, or white with black spots.

- Shiny Leaves.

- Seeds interior pods which might be huge and red.

- Roots which can be very large, fleshy, and broaden from a vital thing.

However, a few poisonous wild plants don't have any of the above tendencies. It is as plenty as you to analyze which plants are toxic.

The Dangers of Eating Wild Plants

It is easy to be tempted into consuming wild flora, particularly in case you are misplaced for days without meals. Some wild plants taste so scrumptious they're tough to face up to. However, if some element tastes right, there might be a reason for the flavor. It may be due to the truth the plant is poisonous, and you'll come to be with a stomachache or worse. Poisonous plants won't be deadly in most cases, but they're able to however make you pretty depressing for some days.

Although some poisonous wild flowers might not cause any issues when ingested, they may create troubles if the plant is treated and the attention location touched. For instance, poison ivy motives pores and pores and skin contamination which can change into blisters. Some toxic wild flora are pretty volatile and can kill you if ingested. These flowers normally taste highly bitter and purpose swelling of the mouth, throat, and tongue. Even in case you spit it out proper now, minute debris would possibly notwithstanding the truth that be swallowed and motive essential damage.

One of the most vital topics to don't forget with poisonous plant life is that they're often in the equal circle of relatives as safe to eat wild plant life. For instance, poison hemlock is in the parsley circle of relatives. Water Hemlock is likewise toxic, and it's inside the carrot family with different secure to consume wild vegetation like Queen Anne's Lace, ground cherries, and cow parsnip. It technique if you apprehend what one plant looks as if, it is simple to pick out out precise members of the same family.

Wild Plants That Can Cause Problems

The following list information some commonplace wild plant life that could purpose extended-term troubles. As you have a look at the listing, keep in thoughts that lots of these flora reason troubles at the same time as eaten or maybe touched.

1. Poison Hemlock

This plant is not unusual in moist, marshy areas. It has a crimson-noticed stem that smells like parsnip or celery while crushed. It can develop to approximately 6 feet tall and has umbrella-customary clusters of small, white flora. This as a substitute toxic plant is inside the equal own family as carrots, parsley, parsnip, and Queen Anne's Lace. The consequences of poison hemlock are considered cumulative, that means you may undergo for days or maybe weeks if sufficient is fed on. The effects embody prolonged salivation, vomiting, paralysis, and probably convulsions. The remedy for this plant is often a mixture of alcohol, milk, and sugar or salt.

2. Water Hemlock

This is a wetland plant that could develop up to 10 feet tall. It has a pink-noticed stem and umbrella-formed clusters of small, white plants. The foliage is extraordinarily toxic and causes extreme poisoning with signs like superior salivation, vomiting, and belly ache. This plant is taken into consideration most of the most violent poisons in North America. If you have to ingest or even contact this plant, are trying to find for right away medical help.

3. Pokeweed

Pokeweed is one of the most toxic plants generally positioned in North America. It causes nausea, vomiting, and diarrhea if ingested. The signs and symptoms and signs and symptoms commonly appear round 10 hours after ingestion and usually closing for two to 5 days. If you eat pokeweed, you need to make yourself throw as an entire lot as reduce the amount ate up.

Symptoms of pokeweed poisoning are fatigue, drowsiness, headache, belly ache or tenderness, sweating, lack of urge for food, thirst, and normally feeling unwell. Many human beings observe the flavor to asparagus. It is fantastic no longer to eat pokeweed in any respect, however it's miles top notch to boil the plant for 10 to 20 minutes if you do. Boiling the plant for first-rate 5 mins isn't always enough for steady consuming.

4. Poison Ivy

Poison ivy is relatively commonplace. It has smooth, three-detail leaves and clusters of small greenish-white plant life. The plant grows to about three ft tall. It is usually located in wooded regions and at the rims of forests. The poison ivy plant is notable regarded for an itchy, pink rash that commonly appears within 24 hours of contact. The rash is resulting from the plant's oil (urushiol) that is particularly poisonous. The rash may be slight or substantially painful and can consist of blisters. You can find numerous treatments online to cope with poison ivy or make a paste of baking soda and water. The high-quality way to prevent the rash is to put on gloves in case you apprehend you will be round poison ivy.

five. Angel Trumpet

This plant is a woody shrub with trumpet-formed white or violet flowers and smooth leaves. It can increase up to ten feet tall and is determined in components of the southern United States. Its consequences encompass sleepiness, lack of urge for food, vomiting, blurred imaginative and prescient, and lack of muscle control. These outcomes commonly appear 30 minutes to 2 hours after ingestion and final for 12 hours or extra. If you ingest angel trumpet, you have to are searching out immediately medical hobby.

6. Mayapple

Mayapple has umbrella-long-established clusters of white flora and smooth, inexperienced fruit. It can develop up to two ft tall and is normally located in rich, wet soil. The complete plant is toxic and motives vomiting, diarrhea, and convulsions if ingested. If mayapple is eaten, someone want to drink hundreds of water and prompt vomiting. These results normally appear half-hour to 2 hours after ingestion and final for 12 hours or extra. The treatment for mayapple poisoning is to drink lots of fluids and administer activated charcoal as speedy as possible.

7. Poison Sumac

Poison sumac is normally observed in wet, marshy regions and is frequently wrong for a harmless bush. It has a dark crimson stem and white or yellow-green flowers. If you touch this plant, it may motive severe signs like burning and swelling pores and skin. Poison sumac left at the pores and skin for a extended length can cause blisters and everlasting scarring. If you're exposed to poison sumac, wash the pores and pores and skin with cleansing cleaning soap and water without delay.

8. Stinging Nettle

The stinging nettle is an herb that causes an itchy, pink rash even as the tiny hairs on its leaves are touched. This plant may be determined in wooded regions or close to water. It typically grows as lots as 4 toes tall and has greenish-white flowers. When utilized in cooking, the stinging nettle can be eaten with none sick consequences. The tremendous way to put together it's miles to boil the plant and use salt or vinegar while ingesting.

9. Wild Licorice

Wild licorice has clusters of small, red plant life and curved branches. It thrives in sunny areas and is most customarily placed in plains or meadows. Wild licorice has the same consequences as ordinary licorice, together with poisoning while it's miles eaten in big portions. This plant is typically positioned in small portions and isn't always normally eaten.

10. Jack-in-the-Pulpit

Jack-in-the-Pulpit is a perennial herb that has a greenish-pink, fleshy stem and darkish berries. It typically grows to about three feet tall and is located in moist woodlands or fields. It is most often mentioned for the top notch purple "pulpit" that appears whilst it plant life inside the springtime. The roots and berries are toxic and can cause vomiting, diarrhea, weak spot, paralysis, impaired vision, hassle speakme or swallowing, and extraordinary heartbeat. If you believe you studied which you have eaten a toxic plant, you need to proper now contact your nearby poison manipulate center.

11. Passion Flower

This is an herb with 10 or extra purple-blue petals surrounding a middle of golden stamens. It usually grows over 3 ft tall, has clean leaves, and may be observed in wet open regions. When consumed in huge portions, passionflower reasons hallucinations, slow heart price, slurred speech, and absence of coordination. If you decided which you have ingested a passionflower, searching out clinical hobby immediately.

12. Foxglove

It is commonly located in rocky or sandy soils, pastures, and open woodlands. It has pink bell-formed plants and clean inexperienced leaves. Foxglove can purpose poisoning if multiple leaf is eaten or you bite at the vegetation. If someone ingests a toxic plant, they have to drink masses of water and prompt vomiting. To cope with foxglove poisoning, an individual want to are searching out medical hobby proper away.

13. Caulophyllum

It is usually discovered in moist woods, thickets, and meadows. It has prolonged, slender inexperienced leaves with clusters of small yellowish flora. When ate up in huge quantities, Caulophyllum motives vomiting, diarrhea, and a loss of urge for food. The excellent manner to understand this plant is thru its "butterfly-like" flower with six petals with pointed suggestions.

14. Dicentra

It is a wildflower that typically grows 1 foot tall and has a greenish-white stem with leaves normal like hearts. It prospers in open woods and has a yellow flower with 4 petals. While most elements of this plant are fit for human intake, they'll be capable of motive digestive problems, in order that they have to no longer be fed on in massive portions. When ate up, Dicentra motives a loss of urge for meals and diarrhea. If this plant is eaten in massive quantities, it can purpose immoderate liver damage.

15.   Trillium

This herb has three huge leaves and a single white flower atop its stem. It normally grows up to 3 ft tall and thrives in wet soil, especially close to streams. Trillium reasons excessive drowsiness even as eaten and should not be consumed. If you bear in mind you studied which you have eaten sufficient trillium to poison yourself, you need to trying to find medical hobby as quick as viable.

It is essential to be careful at the same time as you forage — make sure which you apprehend what each plant seems like and if it's far suitable for consuming or not. The terrific way to avoid poisoning is to make certain you recognize the manner to find out toxic flora and typically put together dinner your food. When eating any form of plant, you want to avoid consuming it in large portions, in particular if it has the slightest suspicion of inflicting poisoning. If you think which you have eaten a toxic plant, you ought to proper now touch your close by poison manage center.

This list does no longer cover all toxic vegetation, so most effective consume plant life that you are 100% positive are healthy to be eaten. Do now not devour any plant in case you do no longer comprehend for effective that it's miles stable. Research and examine the flowers on your location earlier than ingesting any plant life.

When You've Eaten a Poisonous Plant

If you or someone else has eaten a toxic plant, severa important steps have to be taken.

1. If feasible, determine the plant's identity earlier than taking any similarly steps. It does not advocate you want to accumulate and discover the plant your self: virtually stay calm and keep in mind as many statistics approximately the plant as you could, which includes wherein it have become decided, the appearance of its leaves and plant life, and its scent.

2. If the man or woman is experiencing signs, hold them calm and however till emergency help arrives, or you could get them to a health facility. Have them live as nonetheless as possible and do now not let them walk or run.

three. Do no longer result in vomiting or supply any fluids to drink besides advised through the use of poison manipulate experts, emergency responders, or fitness care experts. Nausea and diarrhea can worsen the toxic plants' outcomes, so most effective take those steps in case you are directed to.

4. If signs and symptoms have now not but regarded, keep away from eating or eating some thing and do now not set off vomiting unless knowledgeable with the aid of manner of poison control experts, emergency responders, or fitness care professionals.

5. Call the National Poison Control Center hotline at 1-800-222-1222 proper now for similarly instructions. Have a person call in case you are unable to. This hotline operates 24 hours an afternoon, seven days in keeping with week.

6. Make a be aware about the time you referred to as to report the incident and hold track of all plant additives or samples eaten to help poison manage specialists determine what toxin became ingested and how it want to be handled.

7. Stay calm and do now not panic. Symptoms of poisoning will typically seem in a few hours, and treatment can be sought at that factor.

Preventing Food Poisoning

There are several steps to take to prevent meals poisoning.

1. Always wash your palms earlier than ingesting, making prepared meals, or dealing with meals system. This step is the most essential. Wash your hands with heat water and cleaning soap for at least 20 seconds at the identical time as switching between those obligations. Make extremely good you do now not touch some element after touching doubtlessly inflamed food or surfaces.

2. Rinse all raw veggies, fruit, and meat underneath on foot water to put off any soil-based organisms that would contaminate your food within the path of coaching. Do not eat or use any elements of a plant you can not really find out as stable for consumption.

three. Always prepare dinner dinner your food very well earlier than consuming to kill any risky parasites or micro organism, particularly in regions in which potable water might not be available to cool the meals brief after cooking.

4. If you cannot hold your food above a hundred and forty°F (60°C), do not forget canning in a stress cooker, sun oven, or over an open fireplace in vicinity of packing it uncooked and eating it later.

5. Do no longer devour wild vegetation that would cause digestive issues except you have got boiled them for at least 30 minutes to wreck the toxin in their tissues, roasted above three hundred°F (100 fifty°C) to kill parasites, or fried them in oil at approximately 375°F (one hundred 90°C) till the oil sizzles.

6. Do not eat any wild mushrooms besides you can truly apprehend them as secure for consumption, and even then, most effective devour one kind at a time. If you do now not recognize for positive it's far regular, do no longer eat it.

7. Avoid eating meat that doesn't appearance or fragrance proper to save you eating dangerous bacteria or parasites.

8. If you are not positive that your meals is constant to eat, do no longer serve it to extraordinary humans, and do no longer keep any leftovers for another time. It is lousy enough for one individual want to get sick, and it's miles disastrous for everybody within the institution to grow to be sick.

nine. If you aren't positive that your water deliver is secure to drink, do not use it for consuming or cooking besides you boil it for at least 3 minutes.

10. Always p.C. A first beneficial aid bundle deal and apprehend the way to use it.

Foraging is a amusing and profitable hobby anybody should strive as a minimum as soon as. Although it could be very unstable, you can take numerous steps to prevent meals poisoning. Wash your fingers, handiest eat at-danger plants cooked, and be careful whilst dealing with unknown wild vegetation.

You must understand a way to perceive the flora in your region earlier than foraging for them. If you can not genuinely choose out a plant as regular for consumption, do now not consume it. If you come to be sick after eating a plant or mushroom you accept as true with you studied is probably poisonous, inform someone and trying to find help straight away.

It is likewise crucial to cook dinner your meals thoroughly and p.C. A first useful aid package on the identical time as you exit into the barren region to make sure you are prepared for any injuries or illnesses that may rise up. Doing so will assist prevent meals poisoning and be dealt with if your scenario worsens.

Chapter 5: How To Grow Your Own Edible Wild Plants

There are many motives for deciding on to lawn with wholesome for human consumption wild flowers. You can rid your outside of pesky weeds, get one step toward self-sufficiency, or clearly keep cash on groceries. With the added benefits of splendor and peace, an safe to devour wild plant garden can be cherished through the use of every age. Evaluate the splendid vicinity on your lawn. If you live in excessive warmness or bloodless climate, keep in mind constructing an indoor healthy for human intake wild plant garden. A sunny, south-going thru windowsill is a excellent place to begin a whole lot of your plants indoors from seed before transplanting outdoor. Other possible locations are on a porch or patio.

It is beneficial to plot your lawn for ease of get right of entry to, so that you'll need to area taller vegetation in the direction of the decrease returned. It is also beneficial to have a fence or trellis in area for vines a pegboard for smaller plant life like peas and beans, sunflowers, or exceptional trekking flowers. This financial ruin covers records on planting from seed, transplanting from interior to outdoors, and retaining a plant's health.

Choosing the Correct Location

The maximum important issue to recollect even as planting your fit for human consumption wild plants is daylight hours. The majority of your flowers will want at the least 6 hours of direct solar constant with day to live on. Some flowers, including tomatoes, peppers, and eggplants, want even more direct daytime. The top notch manner to tell in case your place is appropriate for the plant is through putting your hand in the the front of it at midday. The plant will probable thrive if your shadow is greater than your hand.

If your place does now not get enough sunlight, go with the flow your plant indoors to an area that receives masses of mild. However, your plant will now not boom as rapid interior and could want commonplace watering. Also, you want to make certain your plant isn't too near a heating or cooling vent, as this can dry out your plant and motive it to die. If you're planning on growing your appropriate for eating wild vegetation in a discipline instead of immediately in the soil, the equal sunlight hours recommendations workout. Use south or west-handling home home windows for maximum plants and east-going thru for plant life like strawberries and raspberries.

Preparing the Soil and Fertilizer

If you're developing your suitable for eating wild plants proper now inside the soil, you ought to make sure it is fertile and mild enough for seeds or seedlings. When tilling the soil, it should enjoy smooth to your contact. If there are rocks or exceptional tough devices, hold in thoughts the use of a lawn fork in preference to a shovel to prepare the soil so that you don't harm your system.

Once your soil is ready, add fertilizer. If you're growing your vegetation right away in the soil, use a fertilizer with an top notch ratio of nitrogen, phosphorus, and potassium. The first variety at the fertilizer bundle deal tells you the percentage of nitrogen. If you're planting in pots, use water-soluble plant food. It is less difficult to control but moreover washes away an lousy lot quicker than granulated fertilizer.

Choosing Seeds or Seedlings

If you live in an area with a short developing season, select flora and greens that best want 3 to four months to develop. If you live in an prolonged developing season place, strive your hand at developing perennial plants. If you need to save money, sprout the seeds your self. This way, you may increase the same fashion of flowers severa times in some unspecified time in the future of a developing season. The only seeds to sprout are alfalfa, broccoli, clover, radish, and mung beans.

When you're ready to plant your seeds or seedlings, make certain the packing containers you're planting them in are deep enough. A proper rule of thumb is to plant your seeds or seedlings about one-and-a-1/2 times deeper than the duration of the seed field they came in. If you need to use pots to your suitable for consuming wild flowers, ensure they have got masses of holes in the backside for proper drainage.

Transplanting Seeds and Seedlings

Once you've decided in which to vicinity your garden and which flora are perfect, it's time to begin planting. For seeds, you'll want to soak them in water for twenty-four hours in advance than planting. It reasons the seed's shell to crack and much less hard for the roots to develop. The water should be room temperature and change the water day by day. After 24 hours, plant your seeds round 2 inches deep to your lawn or a starter pot with correct drainage. Use a seedling mat (determined at maximum domestic improvement shops) underneath your pot to help your seedlings in keeping heat and developing faster.

If you're planting a seedling, make a hole times the intensity of your seedling's region. Place the seedling in and gently pat the soil round it to regular the roots. Be careful no longer to damage the roots at the same time as coping with your seedlings. When planting flowers and greens, additionally ensure to vicinity them successfully to have hundreds of room to increase. If you're the usage of a trellis to your tomatoes, plant them three-four ft apart. Leave approximately 2-3 ft of area among smaller plants.

Tending to Your Plants

Once you've started out your flora, you'll need to do some essential duties each day. The first is checking for weeds. You ought to pull out any weeds growing on your garden as quickly as you phrase them. The 2d challenge is checking the soil for dryness and watering it if essential. Make high-quality now not to over-water your plants and don't allow them to get notably dry. It's tremendous to use a watering can to distribute the water lightly for the duration of your vegetation. The 1/three undertaking is to check for bugs. Slugs are usually a problem when developing meals in the wild, but you may preserve them at bay thru putting down the copper tape. You can discover this at maximum garden stores, and it doesn't have an effect on the plant however maintains slugs and snails away.

It's additionally critical to plant pest-repelling vegetation in and close to your healthy for human consumption wild plant lawn to keep the bugs at bay. Some genuine picks are garlic, catnip, and marigolds. As your vegetation grow massive, you'll want to thin them out once in a while. This way pinching off the smaller or weaker-searching flora at their base till they're about 2 or 3 inches apart. The tremendous time to do that is inside the early morning or past due middle of the night at the equal time because the sun isn't as sturdy. Here are a few useful tips on growing your end result and vegetables:

1. Use Compost

As your plants expand, you may be conscious they start the use of up many vitamins within the soil, specially nitrogen and phosphorus. If you want your lawn to be as healthy as viable, frequently use compost. It keeps the soil rich and gives vitamins essential on your plants' growth. It's important to use handiest a small quantity of compost at the start. When your plants have grown, pass in advance and upload greater compost.

2. Keep the Weeds at Bay

If you want your lawn to thrive, you want to preserve weeds at bay. They'll soak up all the nutrients, water, and daylight hours your flora need to boom. One manner to keep weeds out is with the useful resource of laying plastic over the mattress or border you're planting in. Also, use mulch or possibly newspaper to maintain the weeds beneath control. Be careful not to allow your plastic sheeting live in place for too long. Some gardeners say it traps moisture and effects in fungus and remarkable troubles inside the soil.

3. Make Sure You Use Organic Matter

Another important element of tending your garden is the usage of natural rely. This is something created from decomposed plant or animal depend. Use it to make your soil softer or crumblier. By together with natural depend, you're improving the structure of your soil.

four. Don't Over-Fork Your Soil

Don't overdo it in case you're the use of a tiller to until your lawn. Tilling too much at once will spoil the structure of your soil and depart it dry and hard. Instead, use a rake to loosen the soil a bit at a time. It allows maintain water within the soil better than tilling ever may additionally.

5. Don't Plant Too Early or Too Late

Planting too early may be more volatile on your garden than tilling as it encourages fungus boom and awesome issues for your soil. The exceptional time to plant is during spring or fall while temperatures are milder.

6. Test Your Soil's Nutrients Regularly

You have to check your soil for vitamins within the course of the growing season. Testing your soil makes it less complicated to decide how plenty compost, organic rely, and fertilizer you need to use. It additionally lets in you parent out how wholesome the soil is. You'll realise if there are problems with the soils' pH or mineral content material fabric over time in case you do this.

7. Pick Your Plants at Peak Ripeness

The terrific time to pick out your flowers is after they're ripe. At this point, their seeds will extend, and that they'll flavor better. Fruits and vegetables that aren't picked at top ripeness (or even as the plant isn't pollinated) can take longer to grow or won't even produce fruit the least bit.

8. Harvest Early and Often

It's essential to reap your plant life often. If you don't, they'll end up forced. After some time, they'll no longer produce fruit or seeds in any respect. Since a lot of your vegetation can be perennial or biennial, leaving some behind for the subsequent yr is crucial. This technique putting off factors of your plant in place of the whole thing.

9. Don't Over-Water

Plants need a wonderful amount of water to stay on, however they could't be watered continuously. If you water your plant life an excessive amount of, the roots will rot. Overwatering prevents the vitamins on your soil from being absorbed and makes it extra tough on your plant's shoots and leaves to develop.

10.   Don't Over-Fertilize

Another mistake to avoid is over-fertilizing your flora. This can reason harm to the soil or perhaps kill parts of it. You don't want fertilizer to have a healthful garden, however if you do use it, be careful not to provide your vegetation an excessive amount of at one time. Moreover, the first-rate manner to avoid over-fertilizing your plants is to wait till you're fantastic they need it.

11.   Keep Things Organized

It's vital to keep things prepared to prevent your garden from turning into overgrown. The first-rate manner to do this is to construct a map of your outdoor and use it as a manual for in which you may plant each crop. This manner, you'll understand how many plant life you want and which vicinity dreams which plant life.

12. Optimize Growing Conditions

Once you've commenced out your garden, preserve on top of your plant's desires. Ensuring the soil is wet, the temperature is appropriate for that specific crop, and there's sufficient sun or coloration. If any of those factors are out of sync, it will pressure your plants, causing them to extend troubles like fungal sicknesses.

13. Plant Diverse Crops

It's critical to plant various flora to make certain your lawn can live on any problems. If one crop fails, others will do properly inside the same place. The greater form of plants you've got, the more hazard of an excellent harvest, as a minimum a few times at some stage inside the developing season.

14. Prune Vigorously

Some plants require pruning to deliver the high-quality yield. Plants like hops, grapes, and fruit trees want to be pruned regularly to develop properly. You need to do it first problem within the spring if you need an sizeable crop via autumn.

15.   Store Your Produce Correctly

Last but not least, you need to recognize the manner to shop your produce properly. The fine way is via the usage of preserving them in a cool, dark, and dry area. If they come into touch with air, mildew may develop. This rule moreover applies to mushrooms due to the fact they'll growth brief after harvest inside the event that they're stored in the incorrect situations.

The most essential aspect to keep in mind tending your lawn is to be aware of any pests that pop up. You can studies more about pest control in your healthy to be eaten wild plant lawn by looking the net or asking someone at a gardening middle. If you're surely into gardening, it's an great concept to take a category at a close-by college or network university. However, make sure the trainer is an professional in in form to be eaten wild plant gardening and has non-public experience developing one.

Harvesting Your Plants

Once your in shape to be eaten wild plant life are mature, you'll want to attain them. The wonderful time is within the early morning after any dew has dried and earlier than it receives too hot outside. Make fantastic your harvested flora don't contact every different as you placed them to your bins, or they'll rot. It's crucial to constantly wash any harvested steady to devour wild vegetation thoroughly earlier than ingesting or cooking them.

You need to begin with a spacious backyard or lawn vicinity. Make sure you have had been given enough room to preserve your fit to be eaten wild flora and tools with out them being inside the way of other utensils. It is pleasant no longer to area a few issue else wherein you're planting because it may get inside the manner or kill your flora. The next step is to start gardening and developing your vegetation. But, earlier than you could do that, get all the equipment you want to put together the land.

Edible wild plant life are clean to increase. All you need is a affected character hand, proper tools, water, food scraps (for fertilizing), and time. The trick to growing the excellent plant life is to understand what you're doing. By following the ones tips and recommendations on beginning and cultivating your in shape to be eaten plant lawn, you will be very a fulfillment and revel in the culmination of your difficult work.

All flowers require a similar quantity of care and interest to develop, but you want to apprehend which ones are perennial or biennial and plan your harvest consequently. Don't over-water your plant life thru watering them too much because it motives damage to the soil and might kill elements of the plant. Use fertilizers sparingly due to the reality too much can negatively have an impact on your in shape to be eaten wild plants.

Knowing a way to begin your garden can save you masses of cash at the equal time as additionally offering you with get admission to to unfastened food. But don't overlook about the paintings worried. You'll want to prepare your lawn earlier than you plant and maintain on pinnacle of factors so your lawn can thrive all season lengthy. If you observe those steps, in advance than prolonged, you'll have a flourishing garden wherein the whole thing may be harvested and replanted for years of enjoyment. Remember, if completed well, ingesting wild flora is a brilliant deal extra wholesome than eating shop-offered produce.

Chapter 6: Common Edible Wild Plants And Their Profiles

Most foragers agree that it's far excellent to start foraging stable to devour flora which can be smooth to discover and recognize. This is real for beginners. As previously said, an outstanding place to begin seeking out edibles is your outdoor. With the right season and condition, you could discover an abundance of advantageous edibles. Additionally, not unusual weeds and mushrooms are usually smooth to inform apart from their non-fit to be eaten opposite numbers.

Whether you want initially common edibles or not, this chapter offers a profile of 50 of the maximum not unusual safe to devour wild plant life, in which you can discover them, and a manner to prepare them for safe consumption. So allow's get into them one after the other!

Amaranth Pigweed (Amaranthus)

"Pigweed" is a commonplace call used for lots flora, inclusive of the lambs' quarters. However, this unique pigweed comes from the Amaranthaceae own family, this is why it's also referred to as "Amaranth pigweed." There are as a great deal as sixty amaranth species – all of which variety in edibility.

All amaranths are match for human intake, no matter the rumors that a number of them are deadly. The caveat is that flora that make bigger in a pesticide-sprayed region are possibly to take in the chemical substances, making the vegetation themselves toxic. But this does not endorse that the plant is inherently dangerous.

Additionally, amaranths will be predisposed to accumulate oxalates and nitrates, making them risky for intake. Agricultural fields normally include high nitrate levels from synthetic fertilizer, so do now not forage on houses in which business vegetation are grown.

You can satisfactory devour the seed of amaranth pigweed after cooking. This is due to the reality eating raw inhibits the absorption of the triumphing nutrients.

Taste

As an extended way as maximum vegetables pass, amaranth may want to now not be considered "top shelf." Still, the taste is quite right – even higher than proper. Many human beings agree that lamb's place's pigweed tastes better, however amaranth leaves make foraging well genuinely well worth the strive.

Young amaranth leaves can make a delectable, mildly flavored steamed green, much like steamed spinach. The taste may also moreover range from species to species, with a few being toward bitter than others. Cooked amaranth seeds flavor nutty, and lots of foragers describe the taste as being "earthy."

Nutrition

Nutrition-clever, amaranth is stated synonymously with "amaranth seed" due to the fact most of the to be had vitamins statistics specializes in the seed. However, as a informal forager, you are more likely to access the leaves, in particular due to the fact they require less attempt than the seed.

The veggies appearance just like beet greens, chard, and spinach – which all fall inside the identical circle of relatives. However, amaranth consists of instances as many nutrients as kale and spinach. It moreover has higher Vitamin A and Calcium stages and numerous different essential vitamins and minerals.

Amaranth seed incorporates fewer carbs than white rice and buckwheat, but it has complete protein.

Identification

Amaranth pigweed alternates among oval-shaped and diamond-fashioned leaves that have a tendency to increase up to 6 inches extended. The plant itself grows over six ft tall. Usually, the stem is greenish, however it starts offevolved to expose pink because of the reality the plant matures.

While most pigweed species develop upright, Amaranthus blitoides, moreover known as prostrate pigweed, growth along the ground.

The stems of maximum amaranth pigweed species can be smooth or with nearly invisible hair, other than Amaranthus spinosus, which comes with thorns.

Amaranth vegetation increase in hundreds by means of way of the stem. You can effortlessly apprehend them in fields due to the fact they seem above flowers like cotton. However, the most identifiable characteristic is possibly to be the flower spikes.

Eventually, the plants dry out, and husked seeds emerge to be harvested.

Habitat

North America is the first-rate location to find local and delivered pigweed species. In reality, you may locate at the least one species in each a part of the entire continent. Amaranth pigweed has a tendency to broaden in yards, fields, at the threshold of woods, and in precise disturbed regions.

The associate plant is lamb's quarters – so, anywhere you find out this, you're in all likelihood to find out amaranth as properly. They each increase nicely under similar times.

How/When to Harvest

You need to best acquire more youthful amaranth leaves. These are small to medium-sized, which means that they may be tenderer and richer in vitamins than the larger leaves. Once the seeds start falling off the plant, that is a sign that they'll be prepared to achieve. This usually happens amongst mid and past due summer season.

First, smash the whole flower head and wrap it in a mesh bag or paper. Then, location it in a well-ventilated, in element shady area in which it can dry for at least one week. Or you may located it in a closet in which you've got a dehumidifier.

When the seeds dry up and are ready to split, gently beat the mesh luggage in which you have got them with a stick. Then, use a clear out to dispose of the stricter chaff. After this, you can get right away to cooking or keep away in a smooth discipline, together with a mason jar.

How to Cook/Use

Prepare pigweed leaves inside the same way you prepare your spinach - steaming, stir-frying, or sautéing. Never consume the seeds uncooked – put together dinner in boiling water for as a good deal as half-hour. If you want a porridge-like texture, cook dinner dinner dinner one element amaranth seeds to 3 components boiling water. One-thing seed to one-trouble water will provide you with a thick consistency.

Asparagus (Asparagus officinalis)

Asparagus officinalis is not a wild plant. Instead, it is more feral. The Europeans introduced asparagus to the New World. Now it exists in all the states within the United States and the provinces in Canada. So you may think that this plant should be anywhere inside the location, but it truly isn't always.

Most US states have asparagus and asparagus-free zones. It have become as soon as intricate to recognize which zones to find out asparagus, however now, you can get this statistics if you take a look at the USDA map.

Identification

The specific element is that asparagus is simple to find out. When you come upon one, the number one element you can note is the spear. However, locating the plant itself is the real art work. You stand a higher chance if you look for vintage plants from the previous season.

Asparagus dies lower back after growing above the floor each season because of the reality it's far an herbaceous perennial. It can be up to 6 ft tall, that's pretty excessive, and the leaves appearance ferny, just like dill or fennel.

There are male and woman asparagus, however the girl plants are splendid because of the fact they sooner or later expand purple berries over the ferny foliage. Note that the berries are poisonous, so that you ought to now not devour them.

In past due fall, asparagus turns to canary yellow at the same time as it does all over again. This is not a shade you will locate on most loss of lifestyles vegetation, so it's far each one of a kind way to find out this plant in fall.

Habitat

Asparagus grows in regions with alkaline or saline soil. Although it best grows in locations with full solar or nearly whole sun, it additionally likes to be round lots of moisture – simply enough to take in water with out getting its ft wet.

Check near small timber and briar patches, but do not look in an open timber or wooded region. Asparagus likes to enlarge during the curly dock, wild mustard, hemlock, and tules. Look at the rims of farms and fields, ditches, hedgerows, and fence traces.

How/When to Harvest

Once you discover a mature plant, use your pocket knife to reduce the spears from the floor. In ultra-current, an asparagus plant grows thin spears, then thick ones, and in the end skinny ones over again. If you preserve the plant after cutting, that you have to, you could skip decrease again to attain greater more than one instances. Asparagus multiplies in a appropriate state of affairs.

How to Cook/Use

You are probably already familiar with the steerage of asparagus. Foraged asparagus may be eaten raw or pickled if that's what you make a decision on. If you need to keep the plant, strive blanching its spears in a large bowl of boiling water with salt for up to 3 mins. Then, placed the spears in a bowl of iced water to marvel them. Finally, you could dry, seal, and freeze for your refrigerator.

Burdock (Arctium lappa)

Burdock is a standard herbal treatment, which can also feature a wild match to be eaten. The roots, leaves, and stalks of burdock are some of the tastiest wild edibles you may ever have in case you put together them the right way. Chances are you are already familiar with burdock due to the fact the plant with the disturbing burrs that get connected for your clothes.

Identification

If you've got ever walked via manner of a burdock plant, you recognize the manner it were given its name. At the give up of its 2nd season, the plant shoots up a quite tall flower stalk, and the flowers turn out to be tiny balls of spikes that wait that allows you to pass if you want to hold on along side your garments.

Most human beings hate burdock for this, particularly because of the fact the spikes are difficult to remove, and that they attach themselves to nearly the entirety. However, in case you hate this plant, finding out it's suitable for ingesting could in all likelihood trade your mind.

Habitat

Burdock grows alongside the rims of strolling paths in which animals and people can select up their seeds and deposit them in exceptional locations. So, you can genuinely take a look at for this plant in any recognised taking walks paths round you – that is in which you're possibly to peer them.

If you ever mistakenly acquire and deposit burdock seeds to your lawn or outside, do now not be surprised to find out them littering the place the following season.

How/When to Harvest

Burdock is a biennial plant, that means that you can harvest it at one-of-a-kind instances of the year. The root is typically harvested in fall on the give up of its first three hundred and sixty five days or in its 2d spring right in advance than it shoots up. Then, in the 2d boom one year, burdock sends up the tall flower stalk. And thru the prevent of that 12 months, the sticky burr balls begin appearing – after which the plant dries out and dies once more.

Even even though you have to typically harvest it within the first fall or second spring, the vegetation growth lower decrease back brief. You can find out sufficient root materials to accumulate in only some months of boom. Generally, burdock turns into too late to achieve as soon as it does to seed, like numerous biennials. Every a part of the plant will become uneatable as soon as the plant life seem.

How to Cook/Use

First, you want to peel the flower stalks and use your pocket knife to cast off the stringy outer layer. Then, you can prepare dinner dinner the remaining stalks by way of way of boiling. Note that the flower stalk need to be peeled instances. The first peel eliminates the outer pores and pores and pores and skin, whilst the second one peel gets rid of the stringy outdoors.

Chanterelle Mushrooms (Cantharellus cibarius)

Also known as golden chanterelles, Chantelle mushrooms are considered the maximum well-known suitable for consuming mushrooms. Foragers love the touchy "mildly peppery" flavor. In addition, the mushrooms typically variety from yellow to deep orange hue, this means that you may without problems spot them in the summer time wooded region.

Identification

Chanterelle mushrooms occasionally have caps which may be as large as five inches, but the common is within the course of two inches. And they'll be typically funnel-long-established, with a wavy look. Other mushrooms normally have spherical, symmetrical caps.

Aside from the intense coloration that calls interest, a few different splendid first-rate of the chanterelle is the fake gills that appearance much like forked and wavy wrinkles. They have sharp edges that run down the stem duration from the cap and function the equal coloration as extraordinary mushroom elements. This mushroom has a few one of a kind splendid characteristic in its fruity aroma, which smells like apricot.

Lookalike

The chanterelle has a poisonous lookalike this is the Jack-o'-lantern mushroom. This species grows on timber while the chanterelle grows from the soil. Also, Jack-o'-lanterns have unforked gills and normally shape clumps on the woods they expand on.

Habitat

Chanterelles commonly seem from past due spring through early fall. They pick out spots with shade, moisture, and loads of natural rely. They also have a mycorrhiza courting with timber which takes a particular period to installation, that means they boom in mature forests. You might not find chanterelles in a wooded location it certainly is been reduce down in at the least five years.

You can locate those mushrooms near poplar, oak, maple, and exceptional hardwoods. But moreover they broaden around white pines.

Other tree species in which you are probable to discover chanterelles are hemlock, bay, and birch. In some regions, they increase round pine wood and fir, so that you do not constantly must go searching hardwoods.

Check for chanterelle mushrooms in low-mendacity damp places and near streams. They have a knack for acting in drainage paths in which the taking walks water can deliver their spores downhill.

When you find a few, are searching for up and down the hill as you are possibly to discover more.

How/When to Harvest

Once you've got got found an amazing spot to harvest your chanterelle, go together with a pointy pocket knife. Be careful as you stroll for the duration of the vicinity to avoid trampling and disrupting the mycelium, responsible for spawning new increase.

Always leave the smaller mushrooms in the returned of even as you discover an secure to consume colony, especially in some unspecified time within the destiny of the rainy season. Go decrease lower again after rain, and you are possibly to look larger mushrooms, so long as you depart them in high-quality form the number one time you go to.

Pull or lessen the chanterelles, but make sure you do not trample on them. Just pull them proper out and throw them inside the foraging basket. Don't harvest mushrooms which is probably too dirty, as those may be quite hard to smooth.

How to Clean

To thoroughly smooth the mushrooms, tear them into halves and use a toothbrush to easy away the embedded dust. Cleaning is the first actual step within the processing and training of any wild mushrooms.

As mushrooms extend, grit often works its manner in the stem. If you do now not damage the stem to easy the dirt internal, you can discover your self chewing on a bit of sand.

How to Use

Chanterelle mushrooms are tremendous additions to stews, soups, and sauces. They additionally pass nicely with factors, wines, and herbs along side:

- Chicken
- Venison
- Garlic
- Tarragon
- Chervil
- Chives
- Ramps
- Beef
- Fish
- Pork

Some specific additives to combine them with embody shallots, onions, veal, eggs, and many others.

Curly Dock (Rumex crispus)

Curly dock is also referred to as the yellow dock, one of the many invasive wild flowers you can discover in North America. The plant is harmful to cattle, sheep, and horses, and its seeds are toxic to chicken. Therefore, it isn't some thing to preserve in your pasture. But, recognize that it isn't always some trouble you may quick cast off for folks who've already got it.

But happily, it would now not absolutely take over your yard as different flora do. Still, it is able to be pretty complex to manipulate in case you do have it in your lawn. An attempt to rid your outside of the yellow dock will in reality result in extra increase.

However, it's far an suitable for eating weed, so that you're higher off treating it as a crop in its early and mild degree, that is the splendid time to devour curly dock.

Identification

The curly dock has leaves with "curly" edges, it's in which it gets its call. Initially, the flowers expand from a basal rosette. But as they grow to be older and begin sending flower stalks up, the leaves start to expand off the stalk alternately.

The hairless leaves are usually inexperienced in coloration, but they once in a while get a tinge of pink as they age. They can increase up to ten inches lengthy and three inches large.

Curly dock leaves are joined to the basal rosette by the use of leaf stems (petioles) enclosed on the node via manner of a skinny sheath of leaf known as the ocrea.

They often start out looking extra like stems than leaves with their period rolled up. At the preliminary diploma, they're slimy due to their mucilaginous us of a. However, the leaves start drying out after they unroll, on the equal time because the stems and sheath continue to be slimy.

The older they get, the extra the stems dry up, and in the end, the sheaths becomes papery.

Habitat

Like most invasive plant species, curly dock has a tendency to develop in disturbed regions. It can be found in fields, pastures, manufacturing websites, roadsides, home outdoor, and so forth. The plant prospers in regions with damp soil, which means that you can also find it near ponds, creeks, springs, and different places near a water frame. Although invasive with the resource of nature, curly dock rarely appears in undisturbed and extra pristine regions.

Taste

Curly dock is part of the buckwheat own family (Polygonaceae), due to this it is associated with vegetation which incorporates rhubarb. It is likewise associated with Rumex acetosa, with the commonplace name garden sorrel. However, it bears no relation to burdock.

Like its near family members sorrel and rhubarb, the dock has a sour flavor – way to the presence of Oxalic acid. The older the dock plant is, the greater bitter and doubtlessly poisonous it turns into.

In cutting-edge, dock leaves aren't delicious enough to be chewed raw – even the younger ones. One aspect that affects the flavor of the dock is the soil on which it grows.

If the leaves are younger enough, they need to have a sour taste a long way from sour, because of this that you may consume them raw. Otherwise, you may need to put together dinner dinner the leaves to neutralize a number of the oxalic acids.

How/When to Harvest

Dock leaves are great harvested in early spring, while the leaves are nevertheless more youthful and moderate, with a mild bitter flavor. Only acquire the smallest leaves, specially the ones nevertheless rolled up, unrolling, or freshly unrolled. Ensure you sample them before on the side of the uncooked leaves in your salad.

Slimy leaves and stems are gentle, which means you can consume them raw. The perception of ingesting slimy leaves might not be all that appetizing, but this is the terrific time to eat them. If you flavor the leaves and they will be too sour, without a doubt discard them. Cooking them to inspire palatability will handiest bring about a slimy mess.

But when you have the time, you could boil them in more than one adjustments of water.

How to Cook/Use

You can also additionally moreover use the leaves in stir-fry dishes, soups, stews, and omelets. Basically, you may use it as an opportunity for spinach within the meals that require it. The leaves may be eaten uncooked, chopped, or cooked in sauces and pasta. Be innovative together with your yellow dock dishes.

Chickweed (Stellaria media)

Common chickweed is also known as iciness weed and is one of the "weed" plant life with diverse makes use of in North America. The plant is scrumptious and nutritious. But, extra importantly, it grows abundantly on the identical time as most plant life aren't thriving.

The plant itself isn't always community to America – it located the European settlers and brief sprawled in the path of the geographical area in no time. Its functionality to hastily dominate a garden mattress is one cause why many gardeners dislike it. Chickweed contains more nutrients and minerals than distinctive vegetation like kale, spinach, and so forth.

Identification

Even for emblem spanking new foragers, chickweed is without issues identifiable. It in no manner grows extra than a couple of inches off the ground, and it has a smooth and stringy appearance. The leaves are small, egg-common with a sharp tip, and that they grow for the length of each distinctive along the stem.

Chickweed plants are tiny and specific, with 5 white petals. The petals have deep clefts that cause them to seem two instances as many. The stem and sepals – the leaves across the flower base – have brilliant visible hairs.

Perhaps the maximum precise function of the chickweed is the single line of hair that runs throughout the stem's duration. It is an excellent way to distinguish common chickweed from its poisonous lookalike, scarlet pimpernel, and the perfect for consuming relative, mouse-ear chickweed.

Habitat

You can locate common chickweed in cool, damp locations with in recent times disturbed soil. Check spherical agricultural fields in late wintry weather, spring, and into the summer time. You may additionally find it in your garden. If what to search for, do now not be surprised if you begin noticing chickweed anywhere you move.

How/When to Harvest

Chickweed is best harvested whilst the temperature is among 35 and 75 stages. This plant actually takes to the air at the same time as the temperature rises to 40 ranges and above. It has a bent to die off in the route of immoderate summer time whilst the temperature hits 90 degrees, but it's going to go decrease returned in fall whilst the weather is mellow all yet again.

Common chickweed grows in superb techniques, counting on the form of soil. If you encounter it solo, you'll discover the stem bent and sprawled throughout the floor. But in a network, the plants stand upright and keep each one-of-a-type together.

Of direction, how the chickweed grows does no longer have an effect on its edibility. However, it is a bargain less difficult to collect the cluster. You can eat each a part of the chickweed that rises above the floor – leaves, stem, flower, and bud.

But you furthermore mght need to be selective at the equal time as harvesting because of the fact you have to only eat the primary one or inches of the stem. After that, it turns into too stringy for the not unusual human palate.

All you need to reap chickweed is a smooth pair of scissors or kitchen shears. Avoid pulling the plants right out of the soil, or you becomes scrapping hundreds of plant fabric.

How to Cook/Use

The remarkable way to devour not unusual chickweed is to snack on it raw. It additionally makes an super addition to salads. However, this isn't the shape of plant that you could hold for long, so you're higher off eating as fast as you can after harvest. You may also moreover dry it for subsequent use.

Chickweed can be used as herbal tea. It has diuretic homes which can be said to be powerful for weight loss and inflammation. To make the tea, simply steep three tbsps. Of chickweed in a cup of boiling water for at the least 5 mins. Then, use a clean out to put off the plant residue earlier than you serve.

Common Dandelion (Taraxacum officinale)

Common dandelions are taken into consideration via the usage of many as a gateway plant into the location of foraging. Not simplest are they clean to understand, but they're moreover tasty and acquainted throughout the us. Like the severa not unusual weeds, dandelions are not close by to America. They followed the Europeans at the tread in their boots.

Identification

The leaves of the not unusual dandelion sprout from a basal rosette, with the quick stems serving as their mounts. They moreover have tooth that problem backward. A single plant may also have as many as ten flora. Each flower grows atop a single stem that sprouts above the leaves. When damaged, the stems leak a white latex-like substance.

The fruits are tiny and yellow, with every containing as a minimum one seed. They develop at the threshold of a cluster of white cotton-like hairs, which appear like puffballs. The roots flow as deep as 18 ft underneath the soil, and they usually shape a completely unique, tall taproot.

Common dandelion's lookalikes are the Cat's ear (similar vegetation but branching stems) and the Sow thistle (tall, furry leaves growing off the stem).

Habitat

One factor approximately commonplace dandelions is that they are not selective about their habitat. They can increase almost anywhere. They pick out entire solar in open areas together with fields, lawns, forest clearings, and distinctive disturbed soils, however they also do nicely in partial coloration.

They can amplify in compacted soil, which allows the soil form considering their roots can assist loosen the dust.

In city regions, you may find dandelions in concrete, mainly among cracks. However, they require damp soil to increase and thrive.

How/When to Harvest

Every part of a common dandelion is wholesome for human intake. However, the leaves are quality harvested earlier than the spring increase and in early fall even as they're younger and gentle. Sure, you may harvest dandelions in the course of the season, but the leaves taste a tremendous deal less bitter even as gathered more youthful.

Those harvested in partial shade moreover taste milder than those who increase in whole sun, but you could dispose of the sour flavor with a few cooking strategies.

Cut the leaves at the lowest with a sharp knife or scissors. Ensure you depart enough leaves to keep the subsequent regrowth. You can keep freshly harvested dandelion leaves in a plastic concern on your fridge for up to two days.

The flora are amassed via sniping at the bottom of the stems. Pick them whilst they may be brightly colored, plump, and attractive on the identical time as they'll be at their high. If you wait too prolonged, the flora can go to seed, making the entire plant unpalatable.

If you may now not be ingesting them right now, placed the stems or plucked flower heads in a jar of water to save you them from final up in advance than you can use them.

You may additionally additionally harvest dandelion roots at any time at a few degree in the developing season. Still, the nice time is in past due fall even as the plant's strength is expended downward.

Find a mature dandelion, hold close the leaves at the very base, and pull right out to uproot the extended taproot from the soil. You might also want a trowel to dig out the plant from compacted earth. Leave bits of root behind to make certain the boom of new flowers.

How to Cook/Use

Young dandelion leaves are a brilliant addition to salads and may be used to make pesto due to their moderate and mild texture. If the leaves are not bitter, you handiest want to wash them and add them to salads. Older leaves are generally bitter, however cooking will eliminate the sour taste.

You can boil or braise both more youthful and older leaves in different dishes the use of various oils, herbs, and spices.

Dandelion flower heads can be used to make tea, or you may eat them uncooked. You can also upload them to salads – they taste sweet and crunchy.

The roots can characteristic a coffee opportunity. You may additionally cook dinner them within the same way you prepare dinner carrots and parsnips.

Elderberry/Elderflower (Sambucus canadensis)

Elderberry is likewise called American elder. It is a shrub that is commonplace in North America. The plant is one among a type with its cream-colored elderflowers, which you could find out on many roadsides in early summer season and past due springs.

Identification

The flower heads are umbel-shaped, and they typically form a cluster. The umbels are approximately six inches in diameter. Some people furthermore call the elderflowers the elder blow. Once the flora are mature, they flip to clusters of tiny dark-pink berries ripened from mid-summer time to early fall.

Elderberry leaves are pointed, toothed, and approximately 3 inches lengthy. The roots, stems, leaves, and bark are toxic, so make certain you do not encompass them whilst making prepared elderflowers. But they will be used for unique medicinal preparations.

Elderberry has a lookalike known as the Hercules' membership – with spherical clusters and thorny stems. The Hercules' club berries are poisonous, so be cautious for them.

Habitat

Elderberries normally make bigger in woodlands, wastelands, scrubs, and hedgerows. You can also moreover additionally find out them along roadsides and out of doors gardens. The seeds unfold thru animal droppings, so check round badger gadgets and rabbit warrens.

How/When to Harvest

Elderberry blooms from June to August at some level in the summer time. First, they flower, after which the berries comply with. Only harvest the quit give up end result while completely matured. This is critical due to the fact the berries are poisonous after they have now not clearly ripened.

Whether ripe or now not, you ought to not eat elderberry uncooked. You want to cook dinner or at least dry them to eliminate the very last poisonous compounds that would upset your digestive tract. Sometimes, the berries appear pink at the same time as the inner is still below-ripe.

Pluck off the clusters of flora at the base and location them right into a mesh bag or basket. Shake off to dislodge hidden insects and insects. The berries must best be harvested later in the growing season, so be selective even as harvesting the flower clusters.

A appropriate rule of thumb is to take a third of each plant.

How to Cook/Use

There are one in every of a kind strategies to apply elderflower. First, you could make tea with the aid of the usage of dipping the dried flowers in boiling water for at the least 10 mins, and then you can steep. You also can use elderflower to make tinctures, salve, soothing eyewash, and syrup. It all is based upon on what you want.

You can preserve the flowers with the useful resource of drying and storing them in airtight jars in a shady spot. Simply arrange the flowers on a mesh display and go away them in a dark area for one week to dry.

When it's scorched, it want to maintain the yellow or white color it had while it modified into glowing. Prevent browning with the aid of retaining in a darkish place sooner or later of the processing.

You can use dried elderflowers as an issue in cakes, pies, cakes, jams, and pancakes. They moreover make a fulfilling addition at the same time as baked with raspberries and strawberries.

Ground Ivy (Glechoma hederacea)

Also known as the Gill-over-the-floor or Creeping Charlie, Ground Ivy is a non-nearby perennial plant not unusual in North America. Like most match for human intake weeds, the European settlers brought it with them. It's no longer a especially well-known wild plant, however it can make an exceptional addition in your wild edibles repertoire.

Identification

Ground Ivy is from the mint own family, so like several members of that circle of relatives, its leaves are organized in the course of from every different, and the stems are square. The leaves are kidney-fashioned with scalloped edges, and they will be approximately 1 inch tall. The leafstalks be part of them to the stem. Usually, they'll be inexperienced in coloration, but they take a purplish tinge in sunny places.

When crushed, the leaves emit a one-of-a-kind scent that is uncommon to individuals of the mint circle of relatives. Most people describe the heady scent as being ugly.

Ground ivy flora are funnel-fashioned with a pink-blue look. They bloom from early spring thru mid-summer season. Usually, no extra than an inch long, they have a tendency to occur in clusters of twos and threes. They develop wherein the leafstalks meet the stem. Each flower has 5 petals.

Habitat

Ground ivy grows in damp, in element shady areas together with the edges of woods. It also loves the complete sun so that you may also moreover find out it in your sunny lawn. In addition, you may discover it in woodlands, hedgerows, grasslands, and wastelands.

The plant creates dense mats above the soil ground by means of the usage of spreading stolons during the floor. Each stolon may be as tall as seven toes, and roots sprout from the leaf nodes to preserve the stolons associated with the ground.

Some of the not unusual lookalikes are Henbit and Purple useless nettle, which might be additionally contributors of the mint own family. Other lookalikes are Common blue-violet and Common mallow. All four lookalikes of Ground Ivy are secure to devour, so you do not must fear.

How/When to Harvest

Ground ivy is first-rate harvested from early spring to mid-autumn, relying to your location.

How to Cook/Use

Younger Ground ivy leaves have a low awareness of volatile oils, which gives mint plant life the minty taste. Therefore, the taste is milder than in older leaves. Younger leaves are also greater slight, because of this they may be the simplest aspect you should take delivery of if you plan on consuming Glechoma.

You can devour the plant uncooked, throw it right right into a salad, or put together dinner like spinach. Due to the effective aroma, you can also use the leaves to flavor your soups and casseroles. In addition, dried ground ivy leaves can be used to make herbal tea.

Hairy Bittercress (Cardamine hirsuta)

Hairy bittercress is a delicious safe to consume whose call is a miles cry from the outstanding. This is wild mustard, so the sour taste leans in the course of spicy-heat in region of sour. Still, it's no longer a heat as one-of-a-type plants from the Brassicaceae own family.

It is known as furry bittercress due to the hairy leaves and stems. The hairs are almost invisible – you may pleasant discover them on extra younger plants in case you look difficult enough with a magnifying glass.

Hairy bittercress is an annual plant that makes seeds in the fall and flowers in Spring. But in hotter climates, it moreover vegetation a incredible deal in advance.

Identification

This is one of the smaller mustards. Therefore, it has compound leaves comprising severa tiny leaflets that expand along aspect the leaf stalks pinnately. The stalks now not often expand extra than six inches, and it's far not frequently multiple leaflet on the tip.

Hairy bittercress' leaves sprout from a basal rosette that appears more quite on the younger plants. The flower stalks are both inexperienced or red, and the leaves' form is unique after they expand from the flower stalks.

The plants are pretty much 2 mm in length, because of this they are tiny. They start off as buds with 4 petals developing on a unmarried, smooth stalk in the center of the basal rosette. The petals are white, in the form of a crucifix, this is why vegetation in the circle of relatives are labeled "cruciferous."

Habitat

This weed thrives in locations with a cooler climate. It dies yet again in summer time while the temperature will boom, in addition to chickweed, a associate plant that you may continuously find nearby furry bittercress. However, it remains usually dormant for the duration of frigid temperatures.

Hairy bittercress likes damp, disturbed soil – the kind you've got got were given in yards and gardens. So take a look at round your backyard or lawn, springs, streams, and other glaringly wet places for this wild fit to be eaten.

How/When to Harvest

Hairy bittercress is considered an invasive plant, so that you want to typically harvest thru pulling the roots from the ground. This is simple to do in lawn soil due to the fact the shape is free. You can also harvest thru slicing the leaf stalks together with your knife or scissors, but go away the flower stalks and seed pods so the plant can regrow.

Unfortunately, the leaves wither pretty rapid. Therefore you ought to not harvest till you're ready to apply the plant.

How to Cook/Eat

Hairy bittercress can be applied in salads, as a garnish, or introduced to a sandwich. The roots are match for human intake, and you may mixture them with vinegar to make a sauce much like horseradish. The plant is an fantastic possibility for micro-veggies, but it's miles an amazing richer supply of nutrients C.

Chapter 7: Common Edible Wild Plants And Their Profiles (Contd.)

Most of the plant life on this chapter are edibles that you could begin accumulating at the primary signal of spring as quickly as iciness fades. The time of the yr at the same time as your wintry climate maintain receives depleted and the brand new three hundred and sixty five days's plants are not prepared is referred to as the "hungry gap."

Fortunately, those wild flora let you fill that hole. Wild veggies are useful in some unspecified time in the future of that thing of the twelve months. Below, you can examine 15 greater common appropriate for eating wild flora that you could accumulate at some point of the traditional hungry hollow.

Cattail (Typha)

When many foragers do not forget a flexible wild plant, cattail might be the number one plant that comes to thoughts. Available in nearly all states of america, this plant boasts a widespread range of makes use of. In exceptional phrases, you could nearly constantly use cattail for some trouble at amazing instances of the one year.

Also referred to as bulrushes, they will be used to make baskets, mats, and packing substances. If you dip the head in oil, you can use it as a torch. The Aboriginal human beings used the roots to make excessive-carb and excessive-protein flours. Its fluffy wool serves as diapers because of its absorbent nature. That is how flexible cattail is.

Identification

The cattail plant has distinguishing talents that make it resultseasily recognizable. First, you have got the tall, stout stalk adorned with a brown cigar-shaped head. Once mature, cattails can develop as tall as eight feet. Next, the leaves are stiff and flat, with a rounded stem popularity erect in the center. Then, you have got got got the cattail flowers, which are separated into male and girl.

The flower head bureaucracy a cylinder form at the stem base, with tiny clusters of male vegetation on the pinnacle and clusters of woman plant life at the bottom. The male set turns yellow on the equal time as loaded with pollen, and you can without hassle collect this in massive quantities.

Habitat

Cattail loves water, because of this you may discover it on lake edges, in streams, swamps, marshes, or anywhere with very wet soil. So test the wet regions, thickets, wet fields, and ditches for your place if you do not have a body of water near you.

How/When to Harvest

The exceptional time to forage the shoots of this plant is in early spring, but the pollen can best be harvested from May to June.

If you harvest on the proper time, you may devour the base of the cattail stem, raw or cooked. To collect, use your pocket knife to reduce the decrease stem and positioned it for your foraging basket or mesh bag. Once you're domestic, easy the stems with as a minimum two to 3 adjustments of water. This is essential.

To acquire greater than certainly the stems, you want to get your hands grimy. So, make certain you've got got your gloves on even as you pass foraging for cattail. The rhizomes are some different suitable for ingesting of the plant that you may collect – they are loaded with starch.

How to Cook/Use

You can use the lower elements of cattail leaves in a salad dressing, while the stems can be boiled or ate up uncooked. The extra youthful plant life may be roasted considering that they are quite smooth. The yellow pollen of a cattail, which has a tendency to appear in mid-summer time, may be used as an aspect for pancakes, growing the dietary rate drastically.

To use cattails for soup or sauce, shake the pollen right proper into a bag and upload a thickener to your stews and soups. Or, you could mix it into flour to make bread. You also can dry and pound the roots to make flour in your meals.

The issue is that cattail can be used to put together dinner in almost any context – you could even add them to stir-fry recipes.

Clover (Trifolium)

Clover is a not unusual in shape to be eaten plant with one of a type names like trefoil, white clover, purple clover, Swedish clover, meadow clover, and lots of others. Unfortunately, the plant is frequently pressured with wood sorrel, and the white and purple clovers are incorrect for each different. Red clover belongs to the legume family, usually used as farm animals food. In assessment, white clover belongs to the pea family.

This plant has a colorful and prolonged information. Traditional Chinese remedy uses it to purify the blood, treatment colds and is even used as incense. Native Americans used it to cope with bronchial issues, in addition to a recuperation salve for burns. Red clover is used explicitly in first rate cultures to deal with respiratory problems, eczema, psoriasis, whooping cough, and so forth.

Identification

You can fast grow to be privy to red clover through the numerous tubular-fashioned plants growing from a crimson flower head. In addition, it has inexperienced leaves with a tinge of inexperienced on the uppermost aspect. That allows call pollinators' interest to the flower head.

White clover has numerous compound leaves that shape from without a doubt three leaflets. The white flora may be observed without difficulty, plus the leaves. The flora additionally shape from the 3 leaflets. The leaflets are finely-toothed, with white patterning in a few vegetation. It is sporadic for 4-leaved clovers to arise.

Habitat

Clovers expand on lawns, yards, strolling paths, waste grounds, roadsides, and seashores with damp soil. You can find out the plant for the duration of the united states and Canada, Australia, Europe, Asia, New Zealand, and some factors of South America.

How/When to Harvest

The excellent time to build up clover is on a sunny day even as the vegetation flavor satisfactory. Also, the first slicing ought to be finished in advance than the primary mid-bloom whilst half of of the plants begin flowering. Then, you can harvest once more at the second mid-bloom, which commonly takes place at some point of summer season. However, that could change depending for your region.

How to Cook/Use

Clover leaves may be tossed uncooked right right into a salad or used to brew tea. But for maximum people, the flower is the greatest a part of the plant. The seeds also are healthy for human consumption – begin with the minutest amount to avoid frightening your stomach. Clover can reason slight or extreme flatulence.

Chicory (Cichorium intybus)

Chicory is a woody perennial plant that grows up to six feet tall. Some of its extremely good names are succory, espresso weed, blue sailors, and every so often, cornflower. It is an herbaceous plant with tasty leaves which are the maximum delicious in spring and autumn. However, the leaves become pretty sour in some unspecified time in the future of summer time because of the warm temperature – however that doesn't have an effect on their edibility.

Identification

Chicory appears scraggly with branching capabilities. It is straightforward to perceive in open fields as it constantly stands on my own, with flowers open on sunny days. The leaves appear like the ones of the dandelion at the bottom, but they will be more spaced and tinier as they development at the stem. They appear pointed and clustered at the bottom of the furry stem. Some leaves even appear on the stem.

Chicory plant life are spherical 4 cm. Extensive. They have a exquisite blue colour and are organized in involucral bract rows. The outer vegetation are quick and sprawling, at the same time as the internal ones are taller and upright. The plant life bloom from July to October.

At its completely grown diploma, the chicory plant can make bigger as immoderate as a hundred 80 cm.

Habitat

Forage for chicory in wastelands, antique fields, weedy parks, grassy areas, and along roadsides. The plant can be determined in maximum of the united states and Canada.

How/When to Harvest

Find clearly lengthy-set up hearts and reduce them above soil diploma. Leave the stumps to sprout once more so that you can harvest extra chicories later. Depending on the sort of chicory you discover during your foray, the "chicons" are typically organized to acquire in underneath four weeks.

How to Cook/Use

All additives of the chicory are wholesome to be eaten. You can eat the younger leaves uncooked as a salad, however hold in mind that they'll be barely sour. Another opportunity is to boil the leaves to function a vegetable. Boiling gets rid of the bitter taste to an extent.

The roots also can be cooked as a vegetable or used as a espresso opportunity. For the latter, you want to roast the roots for so long as crucial until they flip darkish brown, after which you need to pulverize them.

Chicory consists of a few oils that could efficaciously terminate intestinal parasites and worms. All elements, which include the leaves, roots, stems, and flora, incorporate the ones useful oils. Thanks to their presence, you could use chicory to cope with sinus issues, cuts, bruises, gallstones, and moreover improve bowel moves.

This plant moreover makes an extremely good possibility for oats for horses – the protein and fats content cloth are testomony to this.

Field Pennycress (Thalspi vulgaris)

Field pennycress is a member of the mustard circle of relatives that is without difficulty identifiable with the resource of its huge fruit, which has a cabbage-like taste. Like many individuals of the Brassicaceae family, it grows abundantly in its habitat. This wild plant originated from Eurasia, however it's far now large in lots of additives of the area, inclusive of most of america.

Identification

Field pennycress grows as tall as 2 ft while mature. Its oval, flat, and good sized silicula is one among its maximum distinguishing capabilities. When flowering, the decrease leaves at the stem turn yellow, and the plant can however ripen seeds even though there are nearly no leaves. The siliculae continue to be inexperienced for pretty a while. It office work in abundance, and each silicula consists of numerous seeds.

This wild plant has 4 white flora with four petals that can be as long as 4mm. The vegetation expand in dense raceme which gets longer as it a long time. There are six stamens separated into gadgets, with three on both issue of the ovary. It moreover has 4 sepals which bloom from May through July.

The leaves increase in a basal rosette. They are square and can develop as tall as 8cm. The petiole is large-winged. Leaves are lobed or greater or a good deal less toothed, and that they expand alternately.

Habitat

You can find out the sector pennycress in waste grounds, pastures, roadsides, railroads, and special often disturbed regions. The plant can be decided inside the direction of Europe, Canada, the USA, Northern Africa, Great

Britain, Japan, Siberia, and other international locations.

How/When to Harvest

Harvest problem pennycress leaves in advance than the plant starts offevolved flowering to make sure that the leaves are still secure to devour. If you harvest too past due, they will be too sour to consume uncooked, however cooking may additionally moreover take some of the bitterness away. Even the younger leaves in spite of the reality that have a bitter taste that may not be for your taste.

How to Cook/Use

Younger leaves may be cooked or eaten raw, relying on your taste. You can add them in minute quantities to salads and meals like lasagna. You may moreover prepare dinner dinner them in soups and sauces or use them to make herbs. The leaf is quite rich in protein. If ground into powder, you can use the seeds rather for mustard. You additionally have the selection of sprouting the seed to feature on your salad dishes. And, you may get revolutionary and find out one-of-a-kind techniques to make use of this plant inside the kitchen.

Fireweed (Epilobium angustifolium)

The fireweed plant is referred to as Rosebay Willowherb and Great Willow herb in Britain and Canada, respectively. It is a nearby plant that you can find out all over the temperate Northern hemisphere, which encompass some boreal forests. It become given this name because of the reality additionally it is the primary colonizer plant to appear inside the soil, in particular after wooded region fires. Fireweed draws bees and hummingbirds, making it an crucial plant for honey producers. It is a member of the primrose own family.

Identification

First, you need to realize fireweed types: var. Angustifolium and var. Canescens. Although comparable, every type has its private distinguishing abilities. The latter is not as common in North America due to the fact the previous. Plus, it's far huge, more visibly hairy, and its leaves have fantastic venation with an awful lot shorter stalks. Var. Canescens is the best that many humans call the awesome willow herb or rosebay herb.

The stems of the fireweed plant are reddish, simple, and erect – with scattered leaves in alternate patterns. The veins of its leaves are spherical and do no longer prevent at the edges of the leaves. Instead, they form a round loop and be a part of each particular inside the outer margins. The flowers are large, showy, and crimson, every with four petals and stigmas and lance-lengthy-set up leaflets.

When the fireweed plant first appears in spring, it resembles a few poisonous plants from the lily circle of relatives. However, you could emerge as aware about it with the resource of the round leaf venation shape, which is particular to it in a way.

Habitat

Fireweed is adequate in open fields, pastures, and locations with barely acidic soil. The taller variety can be placed on hillsides, open woods, motion banks, and alongside seashores for the ones within the arctic vicinity. It is specially ample in areas that have been over-burned. The seeds stay for the burnt place to reforest after which germinate after a present day burn.

How/When to Harvest

Fireweed can be harvested at each boom diploma because of the reality there is constantly some thing it could be used for. First, acquire the leaves at the same time as they are despite the fact that pointing upward, close to the stem. Then, you may pinch off the leaves to devour like spinach. Be certain to acquire early as growing older makes the plant more fibrous and unpalatable.

How to Cook/Use

Young fireweed leaves and shoots can be delivered to salads, like a vegetable. The sprouts also can function an possibility for asparagus. As they turn out to be older, the shoots turn out to be too fibrous to ingest. You can cut up the stem open and consume the pith raw – that may be a properly supply of Vitamins A and C. The root may also be roasted after being scraped. However, it may flavor bitter.

If you dry the leaves after harvesting and keep them in luggage or jars, you can use them to make tea on occasion. There are distinctive techniques to make use of the secure to consume elements of the fireweed plant, however the ones are some of them.

Garlic Mustard (Alliaria petiolata)

Garlic mustard is a not unusual biennial wild herb from the mustard (Brassicaceae) circle of relatives. Some of its other common names consist of hedge garlic, jack inside the bush, jack within the region, negative man's mustard, and so forth. It is neighborhood in Europe, Central and Western Asia, and North Africa. It grow to be introduced to North America, in which it has become some component of an invasive species. The plant has a taste that could be a blend of mustard and mild garlic.

Identification

The stems of this plant are erect and furry, with big toothy leaves which can be inexperienced in coloration. Sometimes, the leaves appear inside the form of a coronary coronary heart. The vegetation are tiny and white with petals that appear like a crucifix.

Habitat

You will find out this wild stable to consume growing abundantly in the rims of woodlands and hedgerows, alongside roadsides and footpaths, and one-of-a-type disturbed areas.

How/When to Harvest

You can harvest garlic mustard from March through September. It is exceptional collected in early or mid-spring. After it plants and the weather becomes hotter, the plant becomes particularly rank, making it unpalatable.

How to Cook/Use

Finely-chopped garlic mustard leaves may be brought to salads, no matter the truth that sparingly. You may also eat it in cheese sandwiches. Many foragers like to apply this plant to make the broadly acclaimed pesto sauce, but you can moreover pair it with lamb.

Consider including it for your stews, soups, and sauces on the give up of cooking. Otherwise, your meal also can taste sour.

The vegetation may be used as a garnish on your salads, on the identical time as the dried seeds may be your terrible man's mustard. Additionally, its taproot has a subtle horseradish flavor that is going nicely in satisfactory dishes.

Don't forget about that the plant releases a garlic perfume and flavor at the equal time as its leaves are overwhelmed, so it can also alternative for real garlic. You can use it the equal way you operate garlic at the same time as cooking.

Finally, the roots have wasabi notes – the taste can variety from "sweet with a touch warm temperature" to "heat," relying on your place and the area you collected it from.

Sea Lettuce (Ulva lactuca)

Ulva is considered the greenest seaweed that one must ever gather from the shore or sea. Ten species are placed in cool water in nearly all components of the area. Sea lettuce is also known as green seaweed via many.

It is an terrific element for salads and pizza. Like any seaweed, you need to reap it carefully to keep it developing and flourishing, so you can keep having bountiful harvests. The top element is that the wild match to be eaten is easy to forage, and it's filled with useful nutrients.

Some states most effective allow foragers to build up sea lettuce off the rocks in its live shape. They do not allow foraging for one-of-a-kind forms of seaweeds.

Identification

Sea lettuce has colourful green sheets that can be as much as 45cm extended. In its early growth degree, it is mild inexperienced. However, it turns to dark inexperienced at the same time as mature. Gutweed is a common lookalike that many humans confuse for sea lettuce. Still, the standout distinction is that gutweed seems tubular, and sea lettuce does not. Although it can serve the equal cause as sea lettuce, the lookalike is a lot extra hard to clean, so it's far first-rate to keep away from it preferred.

Habitat

Sea lettuce is normally decided clinging to rocks in rockpool edges and one-of-a-type massive green seaweeds within the inter-tidal place, mainly the numerous tide marks. Check for this wild plant on rocks and rock cabinets in areas in which the tides cowl and discover each day.

It specifically prospers in water our our bodies with excessive nutrient degrees – like river mouths close to populated cities or close to ocean outfalls.

How/When to Harvest

The nice time to reap green seaweed is inside the course of low tide. You can use a sharp knife to lessen as lots as you want, but be careful not to damage its attachment to the rock. That is the amazing manner to make certain a quick regrowth. Remember to take a hint proper proper here and there, however now not an excessive amount of from one particular vicinity.

A appropriate rule of thumb is to lessen the better -thirds and leave the final 1/three intact. The high-quality time to benefit is for the duration of spring or early summer season. That is while the flora are huge, and the vitamins are better.

How to Cook/Use

Before the usage of or cooking, make sure that you supply the seaweed a superb wash in at the least 3 rounds of sparkling water. That will take away all critters and sands that can be within the fronds. Sea lettuce is one plant that can be used diversely. It can serve as a vegetable, herb, or seasoning. You may additionally dry and salt it and then upload sesame, but the way you certainly consume it is your desire.

Kelp (Alaria esculenta)

Kelp is every special fit to be eaten seaweed that you could acquire throughout your forays. There are species: Forest Kelp and Oarweed. Another is known as sugar kelp, however all 3 are known as "kelp" or "tangle" based on areas.

Identification

Both Kelp species have a brown to golden brown colour that makes them stick out from extraordinary seaweeds. They are big and round 3m. Long, with flat fronts protruding from the meristem. You can tell the 2 apart via way of the flattened stem, that is specific to the oarweed – it flops limply on the identical time as uncovered. On the alternative hand, woodland kelp has robust stems that remain upright regardless of how low the tides are.

Habitat

You can discover the kelp species in areas below the imply tide line. Get a tide desk so you can observe the excellent way to acquire kelp without wading. Kelp has a tendency to form a dense subaquatic forest at some stage in the coasts of its habitat.

How/When to Harvest

Tides allowing, you may harvest the kelp at maximum times of the yr. But the remarkable time is from March to June. When you harvest, reduce -thirds of the blades above the meristem and make certain you depart the very last one-1/3 to make sure endured regrowth.

How to Cook/Use

Generally, each kelp species are too hard to chunk uncooked or use as a vegetable, however you might be able to in case you harvest the younger, but-to-mature fronds. The splendid manner to devour kelp is to dry after which consume the blades crisply. Or, you can use it as an thing in soups and shares to impart flavor. The large portions can be applied in broths much like the use of bay leaves – go through in mind to remove them earlier than ingesting.

Kelp can add intensity, nutrients, seasoning, and umami on your meals. More importantly, it lets in all the different flavors to your dishes to in truth shine. The stems additionally can be sliced and pickled or used to make lasagna whilst dried.

Lamb's Quarters (Chenopodium album)

Lamb's Quarters has many names – wild spinach, pigweed, goosefoot, fat hen, and so on. It is a common wild in shape to be eaten that you can discover nearly everywhere in North America. And the exquisite problem? It is to be had via the complete developing season. The exceptional versions of its common names allude to the truth that you may discover it in nowadays disturbed low and wet areas.

The call "wild spinach" shows the taste and texture of the plant, that is why it's miles the second maximum generally used name after lamb's quarters.

Identification

Lamb's quarters are unbelievably clean to recognize, way to the awesome powder that covers the entire plant. This nearly-waxy powder is thickest during the growing suggestions, making the plant nearly waterproof. It moreover allows to hold bugs away.

The leaves have a triangular form, with toothy edges with a touch of red-purple hue every now and then. However, for my part, the vegetation range in leaf shapes. Some are pointy and slim, on the same time as others are extra rounded. The common issue is that the leaves from all of the flowers protrude from the stem in a celeb-like shape.

The plant life can increase as tall as 7 ft in a growing season, but they commonly fall amongst 3 and 5 ft.

Habitat

Lamb's vicinity thrives close to rivers, streams, wooded place clearings, fields, wastelands, gardens, and normally disturbed soils. It grows sooner or later of america and Canada.

It is likewise available in Central America, South America, Africa, Europe, the Middle East, and many nations in Asia.

How/When to Harvest

Lamb's quarters are hardy plant life. When you harvest, make certain you select the youngest flowers which have no longer shaped seed heads. Take a couple of scissors and a plastic bag, and lightly reduce the inexperienced clusters off the complete plant.

How to Cook/Use

The leaves, seeds, shoots, and flowers of the lamb's quarters plant are all in shape to be eaten. But the seeds comprise saponin, which may be poisonous while ate up in extra, so keep that during mind. In addition, the plant itself consists of oxalic acid, which means you have to consume it uncooked in the smallest quantities feasible.

Cooking the plant gets rid of the acid. You can use it in salads or as an detail in smoothies and juices. Another manner to cook dinner dinner lamb's quarters is to steam it. You also can add it to soups, stews, and sauteed dishes.

Drying the plant is likewise one way to characteristic it in your meals. Or, if you need, you could blanch, freeze, and store the leaves for destiny makes use of.

Prickly Pear Cactus (Opuntia)

Prickly pear cactus is a wild plant that originated in South America and ultimately decided its way into Mexico and Southern US. There are fifteen species from the Opuntia genus: Opuntia basilaris, Opuntia strigil, Opuntia rufida, Opuntia erinacea, and masses of others.

Identification

The prickly pear cactus is a perennial plant with rounded stems known as phylloclades. The stems are fleshy, flattened, and with one-of-a-type joints. The phylloclades are usually known as the pads – essentially modified stems that serve multiple functions.

The flora of the plants are wholesome for human consumption and supplied in grocery shops as "tuna." The pads are also supplied in shops as "nopalito."

Also, the vegetation bloom amongst April and June. The sun sunglasses rely on the species of Opuntia. It may be orange, yellow, cream, peach, crimson, or a subtle combination of colorings. The plant has no leaves.

Habitat

You can find this wild safe to consume in places with full sun and nicely-tired soil.

How/When to Harvest

Prickly pear cactus is typically organized for harvest from early spring to fall, however it is based totally upon on the cultivar. Don't harvest before they will be completely ripe if you need the sweet taste. Wear your thick pair of gloves whilst amassing this plant.

How to Cook/Use

You can put together dinner dinner the fruit and pads of a prickly pear cactus as they will be each in shape to be eaten. But be cautious at the same time as getting equipped it. You can use the pads to make salads, bread, casseroles, stews, omelets, and tortillas. The cooking possibilities are almost countless. The pads may be boiled, steamed, grilled, or sautéed.

Ramps (Allium tricoccum)

Many foragers do not forget ramps the holy grail of untamed safe to eat plant life. Also known as wild leeks or ramsons, they were a number of the earliest wild flowers recognized to foragers. Ramps are loads of untamed onions with a peculiar taste blend of onion and garlic. You each love or hate the taste because of the fact it can be overwhelming for a few people.

Ramps are well-known because of the truth no special plant can rival their taste. Plus, you may only get them at some point of a short time frame in spring.

Identification

Ramps have more than one sorts. The first is the var. Tricoccum with pink stems and massive leaves. The 2d variety is the var. Burdickii, called white ramps or slim-leaf ramps due to the white stems and slim leaves.

The white ramps have a milder taste and flavor than the red-stemmed ramps. Their leaves and bulbs are also smaller in

Chapter 8: Common Poisonous Plants And Their Profiles

The following are vegetation which could purpose destructive effects such as rashes, burning, swelling, and itching even as you masses as are available in contact with them with out enough safety.

Poison Ivy (Toxicodendron radicans)

Remember the "Leaves of three, depart them be" rule? Well, poison ivy is one plant that rule without a doubt applies to. So so long as you hold this rule in thoughts, you may keep away from contact with this not unusual toxic plant.

Poison ivy usually has a large middle leaf flanked by manner of using two smaller ones on each components. The leaf's shape can variety, however the component is continuously pointy. The leaves start with the color crimson inside the spring, then in the end alternate to inexperienced at some point of summertime. Finally, they flip to yellow or orange inside the fall.

Coming in touch with poison ivy can reason considerably itchy pores and pores and skin that might not depart for a long time. The plant is positioned anywhere in the States besides Alaska, Hawaii, and a few factors of the southwestern drylands.

Poison Oak (Toxicodendron diversilobum)

Poison o.K.Is every special toxic plant bearing similarities to poison ivy. It has a huge center leaf and smaller ones on the edges, making 3 leaves. Its name comes from the leaves, which is probably lobed like that of the all righttree leaves. There are hairs on every sides of the leaves, which generally have a duller colour of inexperienced than poison ivy. Like the latter, poison o.K.Can be discovered all through the united states.

Poison Sumac (Toxicodendron vernix)

This unique poisonous plant grows as a tree. It may be as tall as twenty ft, in particular in swampy regions. Poison sumac has reddish stems, and compound leaves with easy edges. The leaves are neither toothed nor lobed. It isn't always uncommon within the Japanese and southern US.

Giant Hogweed (Heracleum mantegazzianum)

As apparent from the selection, this is one giant plant — the massive hogweed can broaden up to fourteen feet in pinnacle. It sports sports an umbrella-common cluster of flowers with at the least 50 rays in every cluster. The vegetation are white in coloration. Touching the big hogweed's sap can reason burns, blisters, and scarring. The plant is commonplace in New York, Maryland, Oregon, Ohio, Pennsylvania, Washington, Michigan, Vermont, Virginia, Maine, and New Hampshire.

Manchineel (Hippomane mancinella)

The manchineel is considered the area's maximum risky tree for a super reason. First, almost each part of the tree consists of powerful pollution that could harm the human digestive tract. In addition, the acidic sap from the leaves can cause blistering, and inside the worst case, make someone pass blind if it touches the eyes. A tiny bite of its small fruit is probably lethal, plus the smoke from burning the leaves and branches can damage the lungs and eyes. That is how awful the manchineel is.

The tree is a tropical wild tree neighborhood to Florida. It appears much like the apple tree, and is the purpose why many people name it the seaside apple. It is frequently decided in salt-water swampy regions and alongside the coast.

Wild Parsnip (Pastinaca sativa)

Wild parsnip belongs to the parsley own family, which includes celery, carrot, dill, and many others. It tends to broaden as tall as five ft, with yellow vegetation that cluster in an umbrella shape. This plant commonly grows interior the issue, pastures, roadsides, and a bunch of various habitats.

The sap from the plant may want to make your pores and pores and skin hypersensitive to daylight hours. The consequences do no longer normally show immediately one comes in contact with the plant. But in case you spend some time within the solar right after touching a wild parsnip, you are high-quality to have a blistering rash breakout. The plant is giant during North America.

Castor Bean (Ricinus communis)

The castor bean plant consists of some of the most surely taking area pollution regarded to man. The oil is used for medicinal skills, however after the toxin has been removed through processing. Eating the plant uncooked may additionally have lethal results.

The Castor bean plant seems like an decorative shrub with huge call-fashioned leaves and easy plant life that appear in clusters atop the stem. The plant is on the begin from Africa, but it's miles been introduced to North America. As a quit end result, it isn't unusual in the jap and southern elements of the us.

You can discover it growing at the brink of agricultural fields, alongside riverbeds, cultivated fields, and wonderful presently disturbed regions.

Daffodil (Narcissus spp.)

Daffodil is a spring desired, but it is also a commonplace cause of puppy poisoning. No a part of the plant is appropriate for eating for each pets or humans. In addition, eating daffodils can reason intense gastrointestinal issues. But the flower is cute with its showy, yellow and white plants. The plant has diverse habitats across North America.

Water Hemlock (Cicuta douglasii)

In an earlier financial disaster, you in brief discovered approximately water hemlock, moreover known as poison hemlock. It comes from the same family as carrot, parsley, parsnip, and fennel. However, hold in mind this plant has no relation with the hemlock tree. Instead, many humans confuse it with the wild carrot and yarrow flowers, which may be every foraged for medicinal and culinary purposes.

Water hemlock has lacy leaves and easy stems. It can develop as tall as eight toes below the right scenario. The stems are sprinkled with a hint of crimson. Its flowers frequently form an umbrella-shaped cluster that looks much like the wild carrot's lace.

Eating water hemlock can disappointed your digestive tract significantly.

Wild Poinsettia (Euphorbia heterophylla)

Wild poinsettia is known as Fire at the Mountain via many. Its leaves are lobed with superb pink blotches at the better leaves' base. The plant exudes a poisonous milky sap that makes identification much less tough. The sap can get worse your pores and pores and pores and skin and is toxic whilst eaten. This plant is native to South America but is also not unusual within the southern a part of america.

Some remarkable toxic flora to observe out for for your forays are:

- White snakeroot
- Pokeweed
- Jack in the Pulpit
- Iris
- Rosary Pea
- Angel's Trumpet

- Jimson Weed
- Deathly Nightshade
- Larkspur
- Corn Cockle
- Foxglove
- White Baneberry
- Monkshood
- Oleander
- Mountain Laurel
- White Hellebore
- Death Camas

Again, here is what to do in case you come in contact with any poisonous plant:

• Wash your hands, pores and pores and skin, and the thing of contact

• Check for symptoms and signs of itching, blistering, swelling, and rashes

• If the symptoms and signs and symptoms persist, call the Poison Help line or go to the closest emergency scientific middle

- Keep a listing of emergency contacts on your backpack whilst you pass hiking

Follow those guidelines, and you could forage correctly even as steering clear of toxic and lethal wild flora.

Chapter 9: Edible Wild Plants Recipes

These are delicious wild plant recipes to function on your series. They consist of number one dishes recipes, soups recipes, salad, sautéed recipes, beverage recipes, and so forth.

Burdock Brown Rice with Mushrooms

Burdock and rice must make the right combination if you recognize how to use them. These commands will come up with a exquisite Japanese-stimulated dish.

Ingredients:

- 4 large mushrooms
- 1 pretty-sized burdock root
- Cold, salted water
- 1 cup of parboiled rice
- 3 tbsps butter
- 1 carrot (finely grated

Instructions:

1. Cut the 4 mushrooms into smaller portions.

2. Clean the burdock root and decrease it into tiny slivers – then pour within the salted water. Leave there for 5 minutes.

three. Follow the package commands to cook dinner your brown or parboiled rice. If you want, add a tbsp. Of miso to the water in advance than putting the rice.

four. Meanwhile, drain the burdock after 5 mins. Place a frying pan on fireplace and soften the butter in the pan. Stir often and sauté for 5 mins. Then, pour the mushrooms and carrots into the pan. Continue to sauté for at least five mins.

5. When the rice is cooked, stir inside the sautéed greens. Wait for 5 minutes in advance than serving.

Garlic Mustard Pesto

This is a traditional that each forager should attempt in spring – with proper motive. Garlic mustard is the ideal pesto plant as it tastes precisely like its call.

Ingredients:

- ¼ cup of almond, walnut, or pine nuts

- four cups of freshly wiped clean garlic mustard leaves. Remove the big stalks, but leave the skinny stems connected to the leaves.

- 1 cup of sparkling parmesan cheese (finely grated)

- ¼ cup of olive oil (greater virgin)

- ½ tsp salt

Instructions:

1. Use a food processor to pulse the almond or pine nuts more than one times until they shape huge crumbs. Then, upload in the parmesan and garlic mustard leaves. Again, pulse severa times until the leaves are clearly minced and the whole lot is properly combined.

2. Keep pulsing and upload in the olive oil. Just pour sufficient till the combination becomes wet and shiny to your eyes. Add the salt to flavor and maintain pulsing. There, you have your pesto.

three. You can preserve it in an airtight jar or field inside the refrigerator in case you are not using it right away. Make effective it aligns with room temperature earlier than serving. You can freeze the fresh pesto for as masses as every week. After that, it is going to begin losing its taste.

four. The pesto may be used on a sandwich, on pasta, or combined into meatloaf. If you need, you may even devour it instantly from the jar with a spoon.

Stir-Fried Dandelions

This stir-fry dandelion recipe is one of the tastiest methods to utilize the suitable for eating wild flowers in the kitchen.

Ingredients:

- 3 cups of glowing dandelion vegetables
- 1 clove of garlic (finely minced)
- ¼ tsp purple pepper flakes
- 1 tbsp olive oil
- Salt (and pepper) to flavor

Instructions:

1. Pour your glowing dandelions into the sink or a bowl filled with cold water. Leave to soak

for a few minutes to dispose of all dust and debris. Then, use a salad spinner or smooth kitchen napkin to dry them.

2. Pour the extra virgin olive oil into a skillet and warmth over mid-warm temperature. Add within the pepper flakes and garlic. Gently stir to prevent the garlic from browning. Once softened, turn up the warmth a piece and toss the dandelion internal.

three. Gently stir the greens to make sure the oil coats them similarly. Keep stirring and shifting so they all contact the pan's base. The goal is for them to become wilted however no longer soggy or limp. This have to take in to 8 mins.

four. Serve the veggies in a dish and devour. Note that the slightly bitter flavor is an appropriate mixture with pepper. It is a facet dish you may use with any meal. The stir-fried greens can also be used as pizza toppings after you chop them into finer quantities.

five. Toss with parmesan cheese, olive oil, and pasta. Or add to an omelet, frittata, or quiche.

Lamb's Quarters Potato Tots

This recipe is for potato children with cheddar cheese, lamb's location, and hundreds of numerous spices. It is straightforward to make and attractive.

Ingredients:

- 4 quite-sized baking potatoes (finely peeled)
- ½ cup of cheddar cheese (finely grated)
- ½ cup of mozzarella cheese (finely grated)
- ¼ cup of chopped lamb's quarters
- 1 tsp paprika
- 1 tsp garlic powder
- 2 eggs, overwhelmed
- 1 cup of panko crumbs
- Flour
- Peanut oil

Instructions:

1. Boil the potatoes in water and permit them to prepare dinner till clean. Drain and permit to relax for a few minutes. Grate the potatoes in a big bowl and pour within the cheese,

paprika, garlic powder, lamb's quarters, and salt and pepper to flavor.

2. Spread a baking sheet for the potatoes. Roll them into small cylinders in advance than setting them on the baking sheet. Then, start dusting each potato in flour, egg, and panko crumbs one after the opposite. Note that the flour, eggs, and crumbs must be in separate bowls.

3. Place each potato tot on the baking sheets till you end all. Then, positioned the sheet inner your refrigerator for 20 to 1/2-hour to allow the combinations to settle in nicely. After this, prepare dinner the kids for your deep fryer till every one is golden brown on all elements. This shouldn't take greater than 5 minutes.

4. Drain the soil and serve with vinegar aioli. The lamb's quarters can be replaced with every extraordinary appropriate for consuming wild veggies which include dandelion, curly dock, Plantago, and lots of others.

Wild Sorrel and Onion Tart

Anyone who loves quiche will love this delicious wood sorrel and wild onion tart with cheese. The meal may be prepped and cooked in much less than an hour.

Ingredients:

- 1 big inexperienced onion, finely chopped
- 2 tbsps butter
- ¼ tsp salt
- 2 cups of freshly picked sorrel leaves
- 1 tbsp flour
- 2 eggs
- Pepper
- 1 cup of heavy cream
- 3 damaged up feta cheese
- 1 pre-baked pie shell

Instructions:

1. Sauté the inexperienced onions in butter on medium-low heat till they may be easy and

clean. Add the salt and flour, and stir in properly.

2. Add the sorrel leaves and preserve stirring for 1 to two mins. Then, take away the pan from warmth. Take a big bowl and whisk the cream, eggs, and pepper together. Add the sorrel mixture into the bowl and set up 3-quarters of the feta cheese.

three. Pour the mixture into the baked pie shell and gently sprinkle the leftover cheese over the top. Put within the oven and permit it bake for forty minutes. Poke a knife inside the middle, and do now not prevent baking until the knife comes out smooth.

four. Allow the meal to relax for 10 mins in advance than serving.

Next, allow's take a look at the salad and Sautee recipes for steady to consume wild flora.

Buttered Chickweed

Combine the freshness of the chickweed with the goodness of inexperienced onions, and you can have this delectable element dish at your dinner desk.

Ingredients:

- 2 cups of chopped chickweed
- 1 finely chopped inexperienced onion
- Butter
- Salt
- Pepper

Instructions:

1. Clean the chickweed very well in bloodless water. Then, pour into boiling salted water. Let it put together dinner for three minutes in advance than you drain the water. You can maintain the liquid to make tea or rice later.

2. Melt a tiny quantity of butter for your stir-fry pan. Pour within the onions and sauté till it is easy and translucent. Then, pour within the chickweed. Add salt, pepper, and special spices that you like. Continue to sauté for 1 to 2 minutes.

three. Serve for 2 human beings.

Plantain Salad

This is a clean summertime salad recipe that everybody to your circle of relatives should like. It packs such a variety of nutrients which might be remarkable for your frame.

Ingredients:

- 2 cups of untamed plantain leaves, finely chopped
- 1 can of chickpeas, properly-tired
- ½ cup of finely chopped cabbage
- 1 finely chopped celery stalk
- 1 finely chopped garlic clove
- 1/eight cup of olive oil
- 1/eight cup of vinegar
- 1 tsp salt

www.ingramcontent.com/pod-product-compliance
Lightning Source LLC
Chambersburg PA
CBHW050402120526
44590CB00015B/1798